STRUCTURAL SAFETY AND ITS QUALITY ASSURANCE

SPONSORED BY
Committee 9A/10 of the Council on Tall Buildings and Urban Habitat (CTBUH)

The Structural Engineering Institute of the American Society of Civil Engineers

EDITED BY
Bruce R. Ellingwood, Ph.D., P.E.
Jun Kanda, Ph

Published by the American Society of Civil Engineers

Library of Congress Cataloging-in-Publication Data

Structural safety and its quality assurance / sponsored by Committee 9A/10 of the Council on Tall Buildings and Urban Habitat (CTBUH) and the Structural Engineering Institute of the American Society of Civil Engineers ; edited by Bruce R. Ellingwood, Jun Kanda.
p. cm.
Includes bibliographical references and index.
ISBN 0-7844-0816-5
1. Structural stability. 2. Tall buildings. 3. Reliability (Engineering) I. Ellingwood, Bruce R. II. Kanda, Jun. III. Council on Tall Buildings and Urban Habitat. Committee 9A/10.

TH845.S75 2005
624.1'71--dc22 2005053573

American Society of Civil Engineers
1801 Alexander Bell Drive
Reston, Virginia, 20191-4400

www.pubs.asce.org

ISBN 0-7844-0816-5
Manufactured in the United States of America.

Foreword

The structural engineering profession plays a key role in the assurance of safe and serviceable building performance. While structural design codes and standards provide the foundation of good engineering practice, the responsibility for proper interpretation and implementation rests with the structural engineer. The safety level that is inherent in any code or standard represents a value judgment on the part of the code-writers (and, by inference the engineering profession and the citizenry) on the question, "How safe is safe enough?" In the field of building construction, this value judgment is based on historical experience. In a time where technology evolved relatively slowly, this approach to structural safety assurance generally was adequate.

However, codes are not a complete guarantee of safety. It came as a surprise to many that numerous buildings and other structures that had been designed to code performed poorly in the views of their owners and occupants in the Northridge Earthquake. One might argue that competent design should take care of the safety issues addressed explicitly in prescriptive codes. On the other hand, structural engineers must pay particular attention to matters that are not covered explicitly in the code or where current prescriptive code provisions may be insufficient. This is especially important for large or monumental structures, including tall buildings. Recent concerns regarding structural behavior under severe fires and building integrity against abnormal loads leading to the possibility of progressive collapse are only two of the many cases in point where structural engineers must apply additional quantitative tools. Buildings meet a fundamental human need for shelter, as well as support the economic infrastructure of a community. The performance of a building can impact a large number of people at once, not only building occupants which may number in the hundreds or thousands but entire communities. Accordingly, the consequences of less than adequate performance in human and economic terms range from severe to catastrophic. Structural engineers practice their art in the public arena. Failures in the built environment invariably are widely publicized, may lead to costly and non-effective remedies, and invariably spark extensive (and expensive) litigation. Expectations of the profession are higher than ever, and penalties for inadequate professional performance are increasingly severe.

Significant advances in the science of structural engineering have revolutionized its practice during the late 20th century. Despite these advances, numerous sources of uncertainty remain in the building process – structural loads and construction material properties are unpredictable; supporting databases are limited; structural systems often cannot be modeled accurately enough to identify performance limits of concern; and the construction process and quality assurance are not perfect. The consequence of uncertainty is risk. Risk is inherent to all human endeavors, and cannot be eliminated entirely. It must be managed through professional practice and informed decision-making in the face of uncertainty. The structural engineering profession is at the forefront in managing risk to the built environment in the public interest. The Council on Tall Buildings and Urban Habitat provides essential technical support to structural engineers in this endeavor.

The Council on Tall Buildings and Urban Habitat (CTBUH) is concerned with the safety, serviceability and economy of tall buildings and their role in modern society. Monographs of the Council facilitate professional exchanges among those involved in all aspects of planning, design, construction, operation and maintenance of tall buildings and serve as a mechanism for international exchange, cooperation and progress. Some recent publications of interest address fire safety, design, codes and special building projects, and tall buildings in the 21st century. The Topical Group on Design Criteria and Loads has responsibility for coordinating council activities in the areas of structural loadings, structural safety and quality control.

An ASCE/IABSE International Conference on Planning and Design of Tall Buildings in 1972 led subsequently to the Monograph (in five volumes) on Planning and Design of Tall Buildings (published in 1978 – 1981), which dealt with safety issues in load modeling, structural analysis and design, and has become a classic reference in the field. In the intervening two decades, the advances in the science and practice of structural engineering noted above have been paralleled by advances in structural reliability theory and applications that provide essential tools to support the structural engineer in assessing uncertainty and managing risk in tall building design and construction. Accordingly, CTBUH Committee 9/10, *Structural Safety and Quality Assurance*, was formed in 1996 to develop a Monograph that would summarize these recent advances in the context of modern international building practices and provide an appraisal of their advantages and limitations as risk management tools.

The Monograph addresses major issues pertinent to meeting the performance goals of tall buildings related to safety, serviceability, durability and economy. Chapters 1 – 3 introduce safety issues in the context of tall building design and summarize the role of building codes and regulations in safety assurance. The current trends toward internationalization of building practices and their significance in terms of ensuring building performance are discussed. Chapters 4 through 9 summarize available quantitative methods for modeling common structural loads, treating the effects of uncertainty in risk-informed building analysis and design of buildings and the structure/foundation system, and assessment of existing structures. Chapters 10 through 12 address the important areas of quality assurance and control, highlighting management techniques that are essential to ensuring that the constructed building is consistent with the building as designed and minimizing the likelihood that human error might derail the design/construction process. Each committee member had responsibility for preparation of a chapter. However, the Monograph has been reviewed by the Committee as a whole and by the Editor, and represents a consensus of the membership.

The field of structural reliability provides a quantitative link between the practice of structural engineering and its social consequences by providing quantitative tools for the management of uncertainty and risk in design and construction. In recent years, the field has grown beyond an academic discipline, and many of these tools have become accessible to professional engineers and decision-makers. The new evolving paradigm of performance-based engineering, with its focus on relating building performance to the needs of the building stakeholders above and beyond the fundamental safety objective of traditional prescriptive codes (which often has been poorly articulated), will cause this trend to accelerate. This Monograph provides a key source of information for the structural engineering profession as it takes the next steps toward implementing these tools in tall building design and construction.

Bruce R. Ellingwood, Editor
College of Engineering Distinguished Professor
Georgia Institute of Technology

Acknowledgments

Chairman
Jun Kanda, University of Tokyo

Vice-Chairman
Karen Chou, Minnesota State University, karen.chou@mnsu.edu

Editor
Bruce Ellingwood, Georgia Institute of Technology,
bruce.ellingwood@ce.gatech.edu

Members
Edmund Booth, Consulting Engineer, Edmund@Booth-seismic.co.uk
Marios Chryssanthopoulos, University of Surry, mkchry@surrey.ac.uk
David Elms, University of Cantaburry, d.elms@civil.canterbury.ac.nz
Dan Frangopol, University of Colorado, dan.frangopol@colorado.edu
Milan Holicky, Czech Technical University, holicky@vc.cvut.cz
Robert Melchers, University of Newcastle, rob.melchers@newcastle.edu.au
Yasuhiro Mori, Nagoya University, yasu@sharaku.nuac.nagoya-u.ac.jp
Jack Pappin, Ove Arup & partners, jack.pappin@arup.com
Lam Pham, CSIRO, Lam.Pham@csiro.au
David Rosowsky, Texas A&M University, rosowsky@tamu.edu
Shunsuke Sugano, Hiroshima University, sugano@hiroshima-u.ac.jp
Ton Vrouwenvelder, TNO, A.Vrouwenvelder@bouw.tno.nl
George Walker, Aon, george.walker@aon.com.au

Contents

Chapter 1

Introduction

By:
Jun Kanda, University of Tokyo
Karen C. Chou, Minnesota State University, Mankato

Buildings are designed and constructed to provide protected space for people to live or work comfortably. The structural safety and the aesthetics are equally important to the functionality of the structure and the comfort of the users. Economical requirements are often the major concerns to achieve this comfort. Environmental aspects are also design issues of buildings particularly those of large scale. Although the height of buildings has been limited by structural safety criteria, engineering technology has been continuously advanced such that buildings have been designed and constructed to greater heights in the last few decades.

Codes and regulations are authorizing documents which provide a required level of safety to buildings through specifications. Usually these are only the minimum requirements. Although these codes and regulations are revised according to the state-of-the-art information at the time of the revision, in many countries these codes and regulations have been used for a long time without revision. Due to some unique issues specific to tall buildings, technical developments specific to tall buildings may not always be reflected in these codes and regulations. The unique features of tall buildings pose additional challenges during the design process. Furthermore, any structural failures would have a higher potential to cause severe consequences since the number of occupancies in tall buildings is significantly higher than in ordinary buildings. Hence, it becomes necessary for the structural professionals involved with tall buildings to keep abreast of the state-of-the-arts development and to utilize the information appropriately.

For tall buildings, issues other than safety must also be considered which may not be the case for ordinary buildings. As buildings get taller, they become more slender. There is higher potential for the slender building to induce motion sickness to its occupants under the wind with loads. A slender building also has a higher potential for mechanical malfunction such as the use of elevator when insufficient damping is available. These are just a couple examples of serviceability failure that tall building designers have to consider. While catastrophes from serviceability failures are rare, economic loss can be tremendous.

Tall buildings such as Kula-Lumpur and Sears Towers were not designed to just meet the clients' need. They also become an icon in society. They may represent an engineering break through, or may reflect societal, cultural, and historical values. Hence, once these buildings are completed, they are expected to be around forever. This poses additional challenges to engineers – durability and maintainability of the structure.

As mentioned earlier, tall buildings impact many users directly and indirectly in society. These buildings usually are located at the business centers of major cities. Any adverse effect on the building would not just affect its occupants it also affects other business establishments. For example, when unexpected maintenance or serviceability failure force half of the occupants of a full building to stay home, the food service businesses near the building will be affected. The impact on these people would be greater as they are usually small business owners. Marginal loss to them financially is high compared to big corporations. Furthermore, with today's global market and business transactions, one can easily imagine the economic loss due to a single incidence of design or serviceability failure. Fatal accidents are no longer the only concern that engineers have to address when designing tall buildings.

For a very tall building, structural safety is still an essential attribute. People expect buildings to be safe against gravity loads, winds, earthquakes, and other possible actions which can cause significant impact on the buildings. However, engineers know that absolute safety is impossible. Nature can produce extraordinary wind speeds or dangerous ground motions. Terrorist acts such as the one that caused the collapse of World Trade Center on September 11, 2001 are design criteria that engineers would not even imagine during the design process.

Professional engineers have an important role of providing an appropriately safe building by utilizing the state-of-the-arts technical tools and information. This monograph attempts to support the professionals in this respect. Students, aiming to be professional engineers, can expect to confront a wide variety of issues on the structural safety of tall buildings.

Structural reliability studies of the last four decades have made big strides in modeling the uncertainties of materials and loads and accounting for engineering modeling imperfection. Results of these studies have been successfully implemented into many design specifications and used widely in many parts of the world. Since these specifications may not always be applicable to tall building design, this monograph presents fundamental concepts of reliability analysis which can be implemented to the engineer's design process or provide starting point for further investigation by engineers into reliability-based design.

Also addressed are other issues associated with tall buildings such as durability in structural safety, quality assurance, quality management, quality controls in construction, and human factors influencing the safety of structures. There are other such issues, as serviceability which is not presented in detail in this monograph for several reasons. One reason is that it is desired that the monograph does not become too voluminous. The second reason is that those topics may interfere with other committee's activities within the Council. Issues regarding designs for un-natural causes such as the collapse of World Trade Center are still in the discussion stage and are not included in this monograph.

This monograph is composed of 11 chapters in addition to the introduction and concluding chapters. The 11 chapters cover the following topics: safety concept, role of codes and regulations, load modeling, reliability analysis, reliability-based design, durability in structural safety and quality assurance, assessment of soils and foundations, assessment of existing structures, quality management of structural design, quality control in construction, and human error. Each chapter is essentially self-contained with an extensive list of references for readers to explore the topic more thoroughly.

Chapter 2

Safety Concepts and Risk Management

By:
David Elms, University of Canterbury, New Zealand

2.1 Introduction

There are many facets to safety. Curiously, safety is best achieved by looking at its converse, at the many ways in which failure can happen (Stevens 1998). This is partly a technical matter: where failure probabilities are low, they are better handled directly rather than as probabilities of survival, which are nearly equal to unity. But it is also a psychological matter. It is better by far to be wary of failure rather than to expect success. Where failures are rare, prudence and wariness are the ultimate guards of safety.

This monograph considers many of the factors ensuring that failures seldom if ever occur in tall buildings. It cannot be comprehensive in detail – that would require far more than a single volume. It does attempt, though, to be comprehensive in its overview, and to bring to the reader's attention matters which might sometimes be forgotten during design, construction and the subsequent monitoring and maintenance of tall buildings.

The discussion is limited to structural safety, and therefore does not deal with serviceability and the business and financial side of creating a building. Neither does it deal with fire safety, even though the maintenance of structural safety in case of fire is a significant issue in building design. Nevertheless some of the tools and approaches dealt with in subsequent chapters can also be adapted for use in fire engineering.

Structural safety is dealt with here as a matter of avoiding structural failure. Historically the primary emphasis has been on avoiding injury and loss of life. This view has changed following the Northridge earthquake, an event causing relatively few casualties but resulting in serious economic loss. Since then there has been a growing realization by structural engineers that design for structural safety must aim at minimization of economic loss as well as protection of people. It is a reflection of what society wants. The primary focus must still be on prevention of harm to people, but now the social consequences of a major disaster must be taken into account. Of course, this relates more to large-scale disasters such as the World Trade Center attacks, or earthquakes and wind storms where many structures fail, rather than to more localized failures such as that of the Hyatt Regency walkways (Marshall et al, 1982).

In addition to being good business practice and a moral issue, structural safety is clearly a social matter. The community demands it. Ever since Hammurabi set out the first crude building codes (Edwards, 1904), society has wanted to require suitably safe structures. Building codes are the means by which society seeks to ensure safety. However, codes in themselves are not a complete guarantee of safety. They are necessary but not sufficient. In a sense, their role is to reduce the need for designers to think of some aspects of structural safety. The correct view of the relationship between codes and design is not that design is essentially a matter of following codes, but rather, that design should principally focus on those matters not covered by codes. This might not be so important with minor structures, but tall buildings are complex, and it becomes imperative that designers should think beyond the code. What is needed is a systemic way of thinking about safety. In part this requirement is covered by quality assurance, but quality assurance, too, has its limitations.

Safety and risk are closely related concepts. Yet they are very different in nature. Risk is quantifiable, but safety, particularly personal safety, generally is not. It is something to be achieved or assured. Matousek (1992) says "Safety is a quality characteristic and should therefore not be restricted to being a mathematical factor." The nature of safety has been discussed elsewhere at some length (Elms, 1999). Suffice to say that in understanding the nature of safety and how it could be approached; the engineer should also be aware of the related concepts of risk, hazard and vulnerability. For tall buildings, safety is achieved within a broader context of risk management by considering the hazards to which the building and its parts might be subjected, and relating this to the vulnerability of the structure. Hazards can be both natural hazards and those deriving from some human cause.

2.2 Overview

Figure 1 gives an overview of the different issues associated with safety. The diagram essentially gives a map which is followed by the rest of this chapter. It also picks up the major themes of the book and shows some of the relationships between them.

Starting with structural safety at the centre and working outwards, the diagram has three main segments: design, construction, and the monitoring and maintenance required during the lifetime of a building. Next comes a band containing some issues common to all three segments: information management, quality assurance, risk management, human error management and quality control. Quality assurance and human error are dealt with in later chapters, while risk management is discussed below.

Outside the common-issue band, each segment contains some of the issues contributing to safety. This chapter is primarily concerned with design, so the two non-design segments contain fewer issues.

Foundations are important to the safety of the structure, but will not be treated further as an independent topic.

Underlying any design exercise is a philosophy, or a broad and strategic way of attaining design goals. To some extent this was touched on above when it was stated that the aim should now be to deal with not only protection of people from harm but also with protection of property and minimization of loss. The attitude of the engineer to code requirements, that design should really start after the code requirements have been met, can also be thought of as philosophical issue in the design process. In a practical situation it is important to understand the underlying methodological framework. An example is the capacity design approach, where undesired failure modes are dominated by more desirable and benign modes. The latter will occur first and ensure that the undesired modes never occur. A fuse is designed into the structure. Another example of a design philosophy is displacement based design for earthquakes.

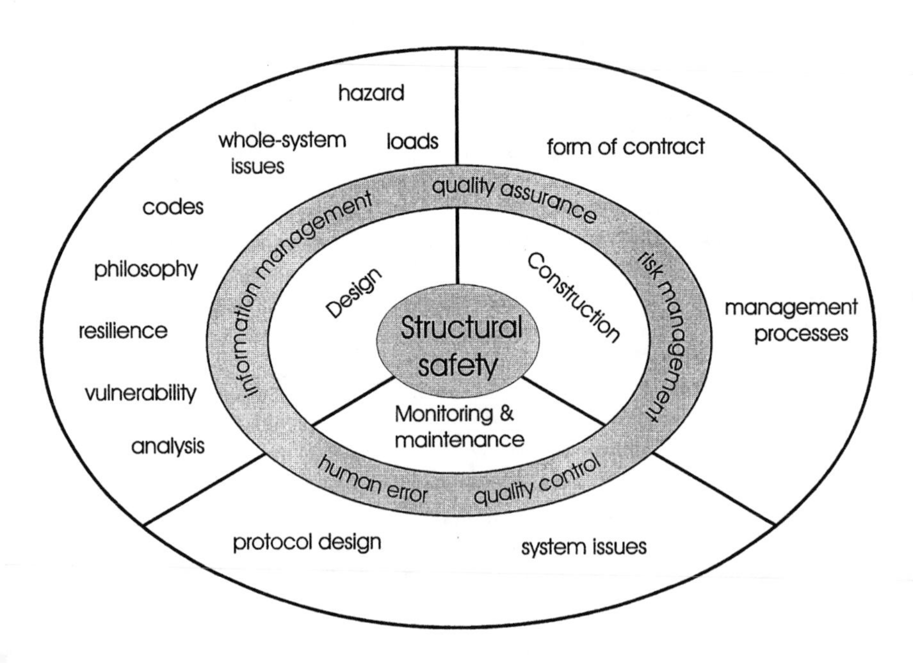

Figure 1 Safety issues

2.3 Design

Generally speaking, society ensures that its structures are safe through the use of building codes. A code, whether prescriptive or performance-based, is essentially a set of rules designed and calibrated to ensure that structures will not fail due to certain causes; that is, due to known and specified demands. An important implication is that there are other failure causes which are not covered by codes, or which are covered inappropriately. For example, most structural codes do not deal well with human factors. They are sometimes assumed to be covered by modifying load and resistance factors, but this assumes that failure will be prevented by simply increasing strength, say. Yet most failures caused by human error have little to do with design strength. A different approach is needed. The underlying question is, to what extent structural design should specifically allow for totally unexpected demands and events.

Most structural codes are prescriptive and state precisely what must be done, what loads must be catered for and so on. However, there is a move towards introducing performance-based alternatives, some of which are now in place. Such codes require a more sophisticated understanding and analysis.

A code for building structures is essentially a means by which a minimum level of safety is achieved no matter to what structure the code is applied. There is a tension between a simpler code which results in a larger variation in safety levels between structures and a more complex and detailed code which achieves a smaller variation in safety by a more precise targeting of specific structural types. By "simpler" and "complex" is meant the degree to which the code disaggregates structures into different types, shapes, materials and usage. Suppose two codes use different ranges of variation to achieve the same minimum level of safety. It follows that the code with the greater variation must produce a higher average safety level over all structures to achieve the same minimum. The simpler code, with its higher variance, therefore leads to a higher overall cost to society (Figure 2). No satisfactory principle has yet been proposed for deciding on the appropriate level of code complexity.

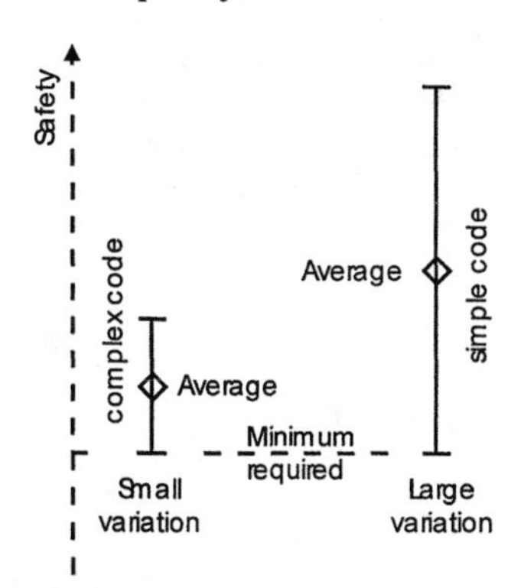

Figure 2 Simpler codes lead to higher average costs

The issue of codes is dealt with more extensively in Chapter 3 and elsewhere (Elms 1999).

Design must take uncertainty into account. It is not just a matter of uncertainty in loads and materials. Structural safety could be threatened by unpredictable events such as sabotage or human errors. For this reason structures must be designed to be robust, to have resilience. That is, they should not fail in a brittle manner, and neither should progressive collapse be possible triggered by a limited failure at one point. The need for this is well understood in shipbuilding and also in the aerospace industry, where the fail safe design philosophy and the introduction of crack stoppers were introduced in the 1950s following major failures such as that of the De Havilland Comet. The requirement to design against progressive collapse has been well-documented in the building industry (Ellingwood and Leyendecker, 1978), but is not always remembered.

One source of unpredictability is error in the design and analysis process. This can be managed in two ways: by ensuring a quality control process is in place, and by managing risk through monitoring secondary indicators of potential trouble.

Quality control requires independent checking procedures to be in place. Beyond this, there are the good quality assurance practices of ensuring that documentation is trackable and so on. Chapters 10 and 11 deals with quality control in design, construction and maintenance. Quality assurance has a significant limitation: it assures the quality of management systems, but not of their content. It will ensure that all procedures have been followed correctly, but it will not normally pick up whether the result of the procedure is correct or appropriate. That is why independent checks are necessary.

The second approach is to look for indicators of proneness to error. An initial set was proposed by Pugsley at an early stage (Pugsley 1973). He suggested that the following factors indicated an increased proneness to structural failure

1. new or unusual materials
2. new or unusual methods of construction
3. new or unusual types of structure
4. experience and organisation of design and construction team
5. research and development background
6. industrial climate
7. financial climate
8. political climate

Pugsley did not intend the list to be definitive. He was discussing the approach, not the detail, and said "Clearly this list could be extended, or each item expanded, or restated under a smaller number of broader heads." The aim was simply to give a framework by which an engineer or a group of engineers could look at a structure or project "and have a good chance of assessing in broad terms its accident proneness."

With a little modification the list of indicators can be made more general. In item 3, for instance, "structure" could be replaced by "product". Item 4 could be expanded to include ongoing management. The phrasing could also be changed to make the items consistent with one another: as the list stands, the presence of some items ("new or unusual materials", for instance) is intended as a bad sign, while others (such as "industrial climate") are neutral. There is also a question of the completeness of the list. An appropriate addition could be "system complexity".

Blockley (1980) has taken the idea further in developing a list of 25 questions to deal with matters "not normally taken into account in structural safety calculations".

In the context of the design process, use of such indicators is related to risk management of the process. The idea is to monitor the indicators and use them as a warning of increased risk within the process. If, for instance, there are novel methods of construction or unusually tight budgetary constraints, then it would be necessary for management to be more than usually wary of error.

Other indicator methods are discussed elsewhere (Elms 1999).

A different approach to dealing with uncertainty is to view the structure as a system and to think of design, analysis, and construction as systemic processes. Safety can then be thought of as a matter of ensuring the soundness or "health" of the relevant systems. System health requires the fulfillment of five criteria (Elms 1998A, B). A system will become unhealthy, or in the limit fail, if there is a failure of one or more of the criteria, namely a failure in

- *Balance*, where some of the elements of the system are too large and others are too small with regard to the system's purpose;
- *Completeness*, where the system has elements missing which are essential to its fulfilment;
- *Cohesion*, where something is wrong with the system's structure, and with the way its elements relate to one another;
- *Consistency*, where elements or connections in the system are inconsistent either with each other or with the system's purpose;
- *Discrimination*, where the various parts of a system and the way they interrelate are unclear, ambiguous or confused.

The five criteria are applicable both to the design itself and to the process of design and its effective management. For example, the balance criterion could be applied to the relative effort spent on analysis, design and project management.

Despite the apparent simplicity of the healthy-system criteria, their application in practice requires care and experience. This is to be expected partly because the approach is subjective and partly because the systems dealt with are complex.

Turning now to the analysis phase of design, there is much to be said for analysis that finds probabilities of failure directly. Though a number of quantitative techniques are available, they are generally suited to specific types of problem and cannot easily be transferred to others. Thus the event- and fault-tree methods widely used for the analysis of failure in chemical plants, nuclear power stations and generally in complex system situations are unsuitable for use in structures. There are a number of ways of categorizing risk and safety problems (Elms 1998C). The key issue here is that, unlike chemical plants, structures are not readily "differentiable". That is, they cannot easily be separated into individual parts whose failure probabilities can be estimated separately. There are, however, techniques especially suited to the tightly coupled systems that are structures. They are the first-order second-moment (FOSM) or first order reliability methods (FORM) discussed in Chapter 5.

First order methods are powerful. They have significant limitations, though. As the number of variables grows, they become increasingly opaque and therefore difficult to check. They are best suited to relatively simple structures, though the interpretation of "simple" is changing with increasing experience and computing power. For some load conditions such as live load, the methods can readily be applied to subsystems of the complete structure, but there are load conditions such as earthquake when the action of a structure can only be considered as a whole. Second-order methods (SORM) are also available.

A more serious question concerns the status of failure probability estimates obtained by the use of such models. If, as is normally the case, human error is excluded, then a major cause of failure is disregarded. Further, such models do not deal with systemic problems, that is, problems arising from the often unexpected interactions of various effects that seem inevitably to arise in large systems. And it is precisely because of the tight coupling of most structures that first-order reliability rather than tree methods have been used.

Hale (1997) talks of three "ages" of safety, referring to approaches to the analysis and management of safety. The first two ages concentrated on technical and human failures respectively. For the third age, "the concern is with complex socio-technical and safety management systems." The history of approaches to structural safety has followed this pattern. Initially the emphasis was on technical issues, concentrating on a better understanding and control of material properties, on loads and on the mechanics of structural behaviour. Next came the recognition that human error had to be taken into account in design, construction, operation and maintenance. Quality assurance spans into the third age, but for the most part structural engineering has yet to find useful paradigms for dealing with threats to safety arising from complex organisational sources. It is unfortunately a characteristic of complex systems that the unexpected will happen.

There is no simple answer. One approach, though, is to assume that reliability analyses will give a partial estimate of reliability, and that therefore other approaches must be used in conjunction with them to give a complete picture of safety. Indicator methods and the healthy-system method discussed above can be seen as complementary to a first-order analysis, and to this can be added quality assurance as an operational approach to ensuring safety. Clearly, such methods do not give precise quantitative assessments of reliability, but in practice too great a precision is both misleading and unnecessary. There are many uncertainties lying between the models of the analyst and the real-life behaviour of a structure. The point is that a mix of methods must be used.

Complex system situations also require a viewpoint significantly different from the cause/effect approach taken by many engineers. It is a matter of looking for patterns, not causes, and of expecting counterintuitive behavior (Blockley and Godfrey 2000).

Nevertheless, first-order methods can be used with confidence when it is a matter of comparative analysis between, say, two competing structural solutions. Precise methods are far more justifiable for a comparative analysis of risk than for an absolute estimate. This, indeed, may be their most appropriate use.

2.4 Construction, and monitoring and maintenance

Many of the remarks in the previous section also apply to the other two sectors of Figure 1, that is, to construction, and to monitoring and maintenance. In these areas, too, uncertainty affects safety, and the means of dealing with uncertainty such as indicator methods and quality assurance are equally as applicable. There are, though, two safety related topics which are specific to both sectors which will be dealt with briefly.

The construction process has a significant effect on the ultimate safety of the constructed building. There are two main factors: the degree of control, and the sharing of organizational risks. The first is a matter of the nature of the control exerted by management, which can be through formal processes such as quality control. These can be thought of as primary controls. Secondary-level control, on the other hand, is to do with the extent to which the work force is encouraged to act responsibly. In essence it is a matter of engendering trust. This is not always easily achieved. A useful book on process-oriented construction-related management has been written by Blockley and Godfrey (2000).

As to the effect on safety of the form of the contract, it can be assumed that where there is antagonism and litigation between the parties, the resulting stresses will affect the safety climate, as Pugsley would see it. A major task of a contract is to organise the distribution of risk between the parties, and when this is seen to be unfair, then there is likely to be trouble. The situation can be helped by the use of newer forms of project contract. Partnering is an agreement added informally to a more traditional form of contract. The idea of partnering is introduced formally into the contract itself in the NEC New Engineering Contract developed by the Institution of Civil Engineers (Barnes 1993). The effect is to transform the contract from a fixed relationship into, essentially, a dynamic management process. FIDIC, too, has developed a modern form of construction contract.

2.5 Risk management

The essence of risk management is control of uncertainty. For a large project, systematic risk management is normally mandatory. It is concerned with the processes of design, construction and maintenance, and seeks to ensure that the risks arising from the uncertainties within the project are identified and treated suitably. Risk management is thus not directly concerned with safety. Structural failure is only one of the risks to be considered. Nevertheless, insofar as structural failure is often due to things going wrong in design or construction, risk management will clearly have a marked effect on safety, and we must consider it here.

In any organisation, risk management is not a separate activity to be applied on top of or in parallel with everything else (Elms 1998D). It must be integrated into the activities and goals of the whole organisation. There are actually two ideas here: permeation and integration. Risk management must thoroughly permeate the organisation and involve management, legal, financial and safety issues as well as design and construction. Moreover, the responsibility for risk management must also permeate the organisation. It is not just the responsibility of a single risk manager or engineer: it must be widespread. However, it is not enough for risk management simply to be distributed across processes and people. It is also necessary for it to be integrated, so that all aspects of risk management work together. They must be connected and structured. The individual aspects of risk management are like the bricks of a house. A house is more than a pile of bricks: the bricks must be ordered and integrated, with each in its right place.

Risk management problems can be classified in three ways: by context, by objective and by usage.

The first classification, by context, can be thought of an axis contrasting physical risk and commercial risk. The two, of course, overlap. An engineering organisation will have to protect itself against both physical and commercial risk. Risk assessment techniques used in the two areas are very different. In part this is because physical risk has more easily described problems. It is easier to put a quantitative measure on the strength of a beam than on the soundness of a business investment. Therefore two sets of expertise have developed, and two bodies of knowledge with little in common between them. The difficulty is that in many cases real problems have no boundaries, and both sets of expertise are needed in the management of engineering projects.

The second classification, by objective, is the distinction between avoiding catastrophe on the one hand, and reducing uncertainty on the other. For both physical and organisational risk management, the aim is generally to avoid catastrophe; that is, to avoid unacceptable or unsustainable loss. Generally the approach to such problems is either to follow a process of first assessing risk and then treating it where necessary to reduce it to an acceptable level or alternatively to design and plan such that the risk is small in the first place. The Risk Management Standard AS/NZS4360 elaborates the first approach (Standards Australia, Standards New Zealand 1999). Another risk management strategy used in catastrophe-avoidance problems is the strategy of risk balancing, which aims to optimise resource use by balancing the risks against the costs of their treatment.

In contrast, for many problems the objective is not to avoid catastrophe but to reduce uncertainty. The idea is to improve predictability. This is the problem faced by the gambler, who needs to know the odds. Farmers, too, need to reduce uncertainty as much as they can, and they can reduce the uncertainty of future income for their crops by selling a crop ahead of time at an agreed price. Futures exchanges are based on this approach. As part of a design exercise, uncertainty in structural behaviour should be reduced as much as possible because, with a lower uncertainty, the design can be less conservative.

The third major classification of risk management problems is by usage. It contrasts the management *of* risk on the one hand with managing *with* risk on the other. The two appear similar, but are in reality quite different. The managing *of* risk is internal, and is concerned generally with reducing the risk faced by an organisation and controlling its acceptable limits. It is primarily a reactive approach.

Managing *with* risk, on the other hand, is a question of moving ahead with prudence, balancing risk against return. The best strategy is seldom the most cautious one.

The three fundamental classifications of risk management tend to overlap in any practical situation. Life is never simple. Particularly, reducing uncertainty tends to go with managing with risk, while avoiding catastrophe aligns with the managing of risk. Together, they give a useful framework for understanding risk problems and for choosing the right approach to risk management.

2.6 Conclusion

In any structure there are many factors contributing to safety. Some are technical, such as structural analyses and knowledge of materials, loads and construction methods. Some are organisational, needing careful management and particularly risk management for large and intricate projects. Some are psychological, relating to human error and its detection and management. Yet others are systemic, arising from the need for many issues to interact and work together for a successful project.

It would be wrong, therefore, to concentrate too much on one contributor to structural safety, or on one means of assessing it or ensuring it. This chapter has sought to provide a brief overview of some of the different facets of structural safety. The main point is this: to ensure safety the engineer has to be aware of a broad range of issues, while at the same time being aware that the world is uncertain. The norm, the expected reality, is that the unexpected will frequently happen.

References

Barnes, M. (1993) *The NEC System,* London, UK: Thomas Telford.

Blockley, D.I. (1980) *The Nature of Structural Design and Safety*, Chichester, UK: Ellis Horwood.

Blockley, D.I. and Godfrey, P. (2000) *Doing it differently: systems for rethinking construction.* London: Thomas Telford.

Edwards, C. (1904) *The Hammurabi Code and the Sinaitic Legislation, with a complete translation of the great Babylonian inscription discovered at Susa.* Watts, London.

Ellingwood, B.R. and Leyendecker, E.V. (1978) "Approaches to design against progressive collapse", *J. Struct Div. ASCE* 104(3):413-423.

Elms, D.G. (1998A) "Indicator approaches for risk management and appraisal", in M.Stewart and R.Melchers (eds) *Integrated Risk Assessment*, Balkema, Rotterdam:53-59.

Elms, D.G. (1998B) "System health approaches for risk management and design", Shiraishi, Shinozuka and Wen (eds), *Structural Safety and Reliability: Proc. ICOSSAR '97, Kyoto, Japan, Nov. 1997*, Balkema, Rotterdam: 271-277.

Elms, D.G. (1998C) "Risk Management - General Issues", in D.G. Elms (ed) *Owning the Future: Integrated Risk Management Principles and Practice*, Centre for Advanced Engineering, New Zealand: 43-62.

Elms, D.G. (ed) (1998D) *Owning the Future: Integrated Risk Management Principles and Practice*, Centre for Advanced Engineering, New Zealand.

Elms, D.G. (1999) "Achieving structural safety: theoretical considerations", *Structural Safety,* 21:311-333.

Hale, A. (1997) "Introduction: the goals of event analysis". In A. Hale, B. Wilpert and M. Freitag, (eds) *After the Event: From Accident to Organisational Learning.* Elsevier Science Ltd., Oxford, UK.

Marshall, R.D., Pfrang, E.O., Leyendecker, E.V., Woodward, K.A., Reed, R.P., Kasen, M.B. and Shives, T.R. (1982) *Investigation of the Kansas City Hyatt Regency walkways collapse.* NBSIR 82-2465. Washington D.C.: National Bureau of Standards.

Matousek, M. (1992) "Quality Assurance", in Blockley, D.I. (Ed), *Engineering Safety*, McGraw-Hill, UK:72-88.

Pugsley, A.G. (1973) "The prediction of proneness to structural accidents". *The Structural Engineer*, 51(6):195-196.

Standards Australia, Standards New Zealand (1999) *Risk Management AS/NZS 4360:1999*, Standards Association of Australia, Strathfield NSW Australia.

Stephens, K. (1998) "Using risk methodology to avoid failure", in D.G. Elms (ed) *Owning the Future: Integrated Risk Management Principles and Practice*, Centre for Advanced Engineering, New Zealand: 303-308.

Chapter 3

ROLES OF REGULATIONS AND STANDARDS

By:
George R. Walker, Aon Re Australia Limited, Reinsurance Brokers, Sydney, Australia
Lam Pham, CSIRO Manufacturing & Infrastructure Technology, Australia

SUMMARY

The nature of society is changing, including the environment in which the commercial activities associated with building and construction take place. To be relevant in the 21st century, structural design codes and standards will need to be international, not national; to be focussed on the customers not the producers of products; to standardise the description and measurement of product performance, not prescribe it; and to be produced in a product based integrated electronic form, not as separate specialist paper documents. They will also need to be separated from local and national building regulations, the principal purpose of which should be to prescribe the minimum levels of performance for the particular community – local, regional or national – for which they are applicable, using the internationally standardised methods of describing and measuring performance. Other issues future regulations and standards will have to deal with include structural robustness under extreme events and sustainable development.

3.1 Introduction

3.1.1 Scope

This chapter looks at the history of building standards and regulations, their different characteristics, modern developments in their format and content, and how they may be changed to suit the changing nature of society.

3.1.2 Definitions

Terms such as building codes, building regulations, design codes and design standards often tend to be used interchangeably in a relatively loose manner. Even when used formally they often have different meanings in different parts of the world as a result of local 1`usage.

In the context of this chapter the following definitions will apply.

Building Regulations: Legal documents issued by national, state or local government specifying requirements that must be met to satisfy the law - includes both administrative provisions and technical requirements.

Building Codes: Documents specifying technical requirements are in a form that can be directly incorporated in building regulations.

Structural design standards: Documents specifying standardised procedures for determining design loads and the resistance of structures to these loads, generally produced by national standards bodies or professional societies, and often incorporated in building codes and regulations by reference, with or without modification.

Design codes (Codes of practice): Documents describing professional good practice for design, evolved from early codes of practice developed by professional bodies. They are non mandatory however, in many countries; structural design codes are synonymous with structural design standards.

3.2 History of Regulations and Standards

3.2.1 Building Regulations

The history of building regulations can be traced back to Babylonian times, where a law stated that if a building collapsed and killed the owner, the builder would be put to death (Edwards, 1904). This would currently be described as a performance based regulation. Modern building regulations, at least in the English speaking world, appear to have had their origin in the Great Fire of London and the Black Death plague in the 17th century. These events led to the imposition of regulations designed to reduce the risk of fires spreading from one building to another, and to reduce the risk to health from poor building construction. These regulations were generally very prescriptive in nature.

The emphasis of building regulations on fire and health has remained a characteristic feature of building regulations until recent times. The incorporation of structural and other requirements has largely occurred in the 20th century, and mostly in it's latter half. More recently, the concept of performance-based regulation has been introduced by the World Trade Organisation as a means for removal of Technical Barriers to Trade (The WTO/Agreement on Technical Barriers to Trade). This will be further discussed in Section 3.4.3.

3.2.2 Standards

Standards have traditionally had two main functions - product assurance and product compatibility. Product assurance means confidence that a product, process or service fulfils specified requirements. Product compatibility means suitability of products, processes or services for use together under specific conditions to fulfil relevant requirements without causing unacceptable interactions (ISO/IEC Guide 2:1991).

By standardising procedures which, if followed, will contribute towards the production of a reliable product, customers are provided with a means for ensuring they get a satisfactory product. Where poor performance of the products can affect human safety or national economic well-being, these standards are often called up in legislation to become regulatory standards.

Structural design standards have their origin in the development of standardised structural design criteria used for the design of major infrastructure works during the nineteenth century when such works were tendered for by private companies on a design and construct basis. In essence they were used to give the client control over the structural safety aspects of the design, and to ensure there was a level playing field in this respect as far as the bidding was concerned.

Initially clients developed their own standards. However through the activities of professional engineering groups wishing to have consistency in design procedures and criteria, national professional standards of good design known as codes of practice were developed. These codes of practice were primarily technical and written for wide application, with many British codes of practice being used by engineers throughout what was then the British Empire. In much of the English speaking world, these were the origins of today's structural design standards.

Product compatibility standards are important where interconnection or the freedom to interchange products is required. In the building industry this has been common in respect of dimensions of some basic building products such as bricks, blocks, steel sections, timber sizes etc.

Historically most standards were prescriptive. This means that they listed a set of prescriptions, which if followed were deemed to produce a reliable product. The level of prescription varied. Some more or less standardised the product (eg. a set of span tables). Others standardised procedures (eg. timber design methodology) thus allowing for more variation and innovation in the product. Among the main examples of the latter was the structural design codes which standardised structural design processes for prescribed materials in terms of specified design loads, and the performance under these loads in terms of the maximum stresses produced in the materials. More recently, the concept of performance standards has been introduced. The aim is to define the performance required in terms of the functional requirements of their users (ISO 6241, 1984). Performance concepts are further discussed in Section 3.4.3.

3.2.3 Merging of Regulations and Standards

In English speaking countries, building regulations and design standards developed independently, and in many parts of the world are still seen to be independent. However in some places, regulation of building became a means by which government could ensure the welfare of all its citizens in respect of living and working conditions. So the codes of practice, formalised as standards, but essentially manuals of good design practice consistent with current knowledge, were increasingly adopted by the regulators as the solutions to regulatory requirements by referencing. Standards may be referenced in regulation, either in whole or in parts, with or without modifications. Once referenced, they become parts of the regulation.

3.3 Characteristics of Standards and Regulations

3.3.1 Regulations

Building regulations are by their very nature the result of a political process, and their application is limited to the area over which the political power to control building practices exists. Historically in English speaking areas of the world, this power did not extend beyond local authorities, so building regulations, where they existed, were very local in their application. This is still the case in many parts of the world, including many parts of the United States. Many of these local authorities did not, and do not, consider building as an area to be regulated, choosing to let social and market forces determine the appropriate level of construction.

Traditionally, the goals of building regulations were to enable the achievement and maintenance of minimum acceptable levels of safety and health in buildings according to community expectation. The roles of building regulations in the protection of properties are limited, in many jurisdictions, to the extent necessary to protect community interest. More recently, there is an increasing demand for the principles of 'sustainable development' to be incorporated as a goal for building regulations. The minimum level of amenity is regulated in some jurisdictions and left to the market force in others. The Economic Commission for Europe has issued a compendium of issues that can be regulated in building construction (ECE, 1996). Most building regulations are prepared with the accepted principle of 'reference to standards'. However the boundary between 'regulations' and 'standards' is not well defined.

3.3.2 Standards

Building standards on the other hand are essentially technical statements that standardise, generally in terms of quality or performance, but sometimes in terms of size or procedure, some activity in relation to building and construction. They can be national or international.

Standards were a product of the industrial age, and designed for an industrial culture in which the producers of products decided what the customer should have. In keeping with this culture they were produced by specialists for specialist users, and the degree of interdependence and integration was very small. Even now most Standards rarely extend beyond single disciplines. They are largely restricted to what can be written on paper in a relatively concise form - to keep the cost down - which means they are limited in the sophistication of the techniques incorporated in them to such forms as tables, charts and simple diagrams and formulae. Advances in information technology have reduced these restrictions enormously.

Standards in many countries are voluntary documents which become mandatory if called up in regulation or in contracts. They are developed and reviewed by technical committees of representatives of interested and/or affected parties. Consensus decision-making principles are generally adopted. For standards that deal with design methodologies, there is the tension between simple and complex provisions, with the former creating more uncertainty, less design effort but possibly more costly solutions. Tiered standards that provide many paths with increasing complexity have been proposed.

For standards that are going to be referenced in regulation, there is the added problem of conforming to regulatory drafting principles, since a referenced standard is considered as part of the regulation. Standards that have their origins as codes of practice are particularly difficult to convert to regulation because the application of many provisions required 'engineering judgement'. There is also the problem of deciding what should be in regulation and what should be in standards. This becomes extremely difficult if the standards are going to be referenced by more than one jurisdiction.

3.4 Modern Developments

3.4.1 Principles of Structural Design

(a) Working Stress Design

Structural design practice developed on the basis of what is now known as working stress or allowable stress design. In this approach design loads were the expected maximum loads during the life of the building – not the maximum probable loads. Working stress structural design codes of practice – and later standards – specified the design loads and the maximum permissible stresses to be used, but little about the basis on which they were derived. Their derivation was assumed to be a role for experts whose consensus on the derivation of the specified values was assumed to represent the best opinion on the requirements for structural safety in the community at the time.

The working stress design approach is still the basis of structural design codes and standards used by many structural engineers around the world. However it has its limitations, a major one being that it does not result in uniformity of structural reliability in terms of safety. Two main factors causing this are the inelastic behaviour of materials before structural failure occurs, and a wide variation in the statistical characteristics of the extreme values of loads of different types. Limit state design was developed to overcome these difficulties.

(b) Limit State Design

The basic approach of limit state design is to define structural design objectives in terms of user needs in relation to structural performance, and ensure the design satisfies each of these. The limiting level of performance associated with each objective is known as the limit state. Initially two primary limit states were identified: the ultimate limit state corresponding to structural failure which is the limit generally associated with structural safety, and the serviceability limit state, corresponding to normal operations of the structure and associated with such characteristics as deflections, vibration, and visual effects such as cracking due to shrinkage or thermal effects which do not affect structural safety (ISO2394, 1998). More recently, attempts have been made to identify other relevant limit states as basis for design. The limit state of restorability, for example, has been received attention in some countries (Okada, 2002).

The traditional approach to limit state design is to assess the design for a number of load combinations in which the loads are specified at some representative values (normally at maximum expected values) with load factors to account for the uncertainties and variability associated with the loads. For environmental loads, some countries have chosen to specify the representative values at the maximum probable levels (representing extreme events) with load factors of 1.0. This approach has certain advantage in dealing with non linear systems or with multiple sets of environmental conditions. While ISO and Euro-Codes recommend that the structural resistance be assessed from characteristic values of material properties and partial factors for material (ISO 2394, EN 1990), some countries have adopted a single factor for resistance (for example AISC, 1986).

Underpinning the limit state design approach is the principles of reliability for structures, in which the basic variables are considered as random variables and are treated with probabilistic procedures (ISO 2394). Various techniques for assessing the probabilities of failure are now available (Ellingwood et al. 1982, Galambos et al. 1982). These tools are useful in 'code calibration' for standard drafting but also can be used in the design of one-off complex structures outside the scope of design standards.

(c) Structural Robustness

Structures are designed to withstand a limited number of extreme events that can be specified. However, they have to be sufficiently robust to withstand unspecified and/or unspecifiable accidents. ISO 2394 states this fundamental requirement as follows: '(Structures) shall not be damaged by events like fire, explosions, impact or consequence of human errors to an extent disproportionate to the original cause'. Building regulations in some countries mention the avoidance of 'progressive collapse' in a similar context. There is little guidance on how to fulfil this requirement apart from some guidance in the British Building Regulation on reducing the sensitivity of building to disproportionate collapse in the event of an accident (HMSO, 1992). The key appears to be the ability of the designer to anticipate the scenarios for these extreme events. Since the events of September

11 2001, this subject has received considerable deliberation but outcomes, in terms of changes to regulations and standards, are still to come.

3.4.2 Internationalisation

A characteristic feature of modern society is the increasing internationalisation of business and commerce. Although the building industry has lagged considerably behind some other industries, it is also being impacted by this change.

There are two ways buildings can be constructed. They can be built in-situ from basic building materials like concrete, steel, bricks, concrete blocks, timber, etc. Or they can be assembled from components manufactured off-site. Currently most buildings, big and small, are constructed in-situ, but there has long been a small proportion of buildings incorporating a significant proportion of precast components, and there is an increasing number of houses being constructed by assembling components manufactured off-site.

In-situ construction is essentially a local activity and provides less opportunity for international trade, or even intra-national trade where distances are reasonably large. On the other hand most facets of construction by assembly of manufactured components are amenable to international trade. The main opportunities are the design and construction management of the building and the supply of less bulky building materials such as steel sections, roof and wall claddings, windows and doors, plumbing and electrical items, and modular built-ins such as wardrobes and shower units.
Where trade by a company is conducted only within the jurisdiction of a single regulatory authority, all the products and services with which it is concerned are subject to the same regulations and associated standards, enhancing productivity and quality assurance, and ensuring a level playing field for trading. Much building activity still falls within this category. However, when companies engage in international trade of services or products then they find themselves having to satisfy different regulations and standards. Since in many cases these regulations and standards are relatively prescriptive and embody local practices, this can be a significant barrier to trade.

The problems created by local regulations and standards in international trading were first recognised in a significant way in Europe because of the close proximity of a number of major countries, each with their own national regulations and standards, and each wanting to expand trade with their neighbours. This led to the development of what was initially known as the European Common Market and now known as the European Union, based on the concept of free trade between member countries. The problem of regulations and standards was recognised early, and this led to the formation of the International Organisation for Standardization (ISO) in the expectation that if standards were made common, uniformity of regulations would follow. ISO, like many of the countries responsible for its creation, is a very bureaucratic organisation and, despite its European influence, is a world body. Getting agreement on International Standards is a slow and cumbersome process. With the creation of the European

Common Market there was an urgent need for common European standards to facilitate free trade across national boundaries. To achieve this, a special European standards organisation was created to achieve this, with European countries agreeing to give its standards higher priority than ISO standards until the European standards had been developed. Furthermore, under what has become known as the Vienna Agreement (1991), it was agreed that any standards developed as European standards would become the first draft ISO standards.

Within the building construction area, a suite of 10 European structural design standards has been drafted, generally known as Euro-codes (Gulvanessian and Driscoll, 2001). These documents establish a set of common technical rules for the design of buildings and civil engineering works for all European member states. Members of the Asia Pacific Economic Cooperation have also realised the importance of having a common international understanding on structural design between owners, operators, users, designers, contractors and construction product manufacturers. Member economies are encouraged to align their national standards with international standards as a first step toward reaching this understanding. (APEC, 1996-2002)

3.4.3 Performance Based Standards and Regulations

(a) Performance-based Regulation

The Performance Concept in Buildings has been a research topic since the 60's (NBS, 1972). The 'directive' for its application in building regulation is in the World Trade Organisation Agreement on Technical Barriers to Trade (WTO, 1994). Clause 2.8 of the Agreement states 'whenever appropriate, Members shall specify technical regulations based on product requirements in terms of performance rather than design or prescriptive characteristics'. Several countries, since then, have investigated the possibilities of performance- (or objective-) based regulation (IRCC, 1998). A number of countries already have or are in course of preparation performance-based building codes such as Canada (NRC, 1995), Australia (ABCB, 1996), New Zealand (BIA, 1993), US (ICC, 2000). Most of these building codes adopted the Nordic model of building regulation (NKB, 1976) with some variations to suit local conditions and practice.

A performance-based regulatory system usually has three components:

- A main document where the objectives and requirements are stated.
- A collection of acceptable solutions, which are adopted by reference, describing methods for complying with the requirements.
- A collection of approved supporting tools to assist with the design and evaluation processes.

For most current systems, these components have not been completely developed. However, the fundamental structures are more or less in place thus allowing the continuing development and improvement of the system.

(b) Performance Standards

In ISO/IEC Directive Part 2 - Methodology for the development of International Standards (1992), Clause 5.2 on the performance approach states: 'Whenever possible, requirements shall be expressed in terms of performance rather than design or descriptive characteristics' (Note the subtle but important difference between WTO 'whenever appropriate' and ISO 'whenever possible'). An important test for 'possibility' is stated in Clause 5.3: 'Whatever the aims.. only such requirements shall be included as can be verified '(Principle of verifiability) and further more '... the stability, reliability or lifetime of a product shall not be specified if no test method is known by means of which compliance with this requirement can be verified in a reasonably short time' (Clause 5.3.2). The above guidance defines the characteristics that performance standards must have.

ISO has also developed a series of International Standards to help with the development of performance standards (ISO 6240-1980, 6241-1984, 7162-1992, 9699-1994). More recently, the need for standardising the description of performance became more urgent, particularly in the building construction area, with increased global trade in construction products and services.

3.4.4 Accountability

Regulations and standards are becoming more comprehensive. This might mislead designers into believing that their duties are limited to demonstrating compliance with the requirements containing in these documents. Most countries, however, place the responsibility of the designers well beyond complying with relevant regulations and standards. For countries with performance-based building codes, the only mandatory parts of the codes are the requirements. Referenced documents are only intended to provide guidance for some of the more common building situations. The documents may be used for demonstrating compliance with the requirements. Their use however is optional and does not absolve the users from legal liabilities.

3.5 The Future

3.5.1 Information Technology

This section discusses the impacts of Information Technology on building construction particularly in relation to Standards, Codes and Regulations. The impacts are discussed under the following headings: (a) communication, (b) information processing, (c) integration and (d) customisation.

(a) Communication

Almost any type of information that can be put on paper can now be communicated almost instantaneously to large numbers of people. This information can also easily be corrected or copied. People at the work face can communicate with each other directly and efficiently, wherever they are located, in another part of the same organisation, or in another organisation.

Standards and regulations of the 21st century will almost certainly be electronic documents circulated and accessed electronically, and integrated through the use of hypertext or more sophisticated referencing systems, so that, to the user, the differences between separate documents used to specify a single product will be seamless.

(b) Information Processing

Large amounts of data can now be recorded and stored in databases from which it can be retrieved effortlessly and almost instantaneously from anywhere in the world, analysed quickly and relatively inexpensively using the most sophisticated analyses available. It is possible to do many things that were previously economically impossible, thus changing the whole approach to many problems.

This should have a big impact on the structural design standards, which have been very constrained by the need to express very complex behaviour in the form of simple formulae and charts which can be readily used manually. What this may lead to is a need to have a number of standardised levels of sophistication of certification, which would be published in conjunction with the technical performance level, and would be associated with different levels of quality assurance. Such a system would enable the customer to differentiate between the technical performance level and the quality of the design and construction process.

(c) Integration

Computer systems can integrate specialist information from wide areas of knowledge effortlessly in very sophisticated ways. Instead of being responsible for the output from an information process, the specialists are increasingly becoming responsible for the input to an information process. The processing itself is undertaken by a system that integrates the knowledge from various specialist areas to produce a result.

Different aspects of Standards and regulations have, in the past, produced independently such as those for fire safety, heating and ventilation, acoustics, lighting, plumbing etc. They will have to be integrated to meet the needs of the producers and consumers who will see these as different aspects of one product - the building.

(d) Customisation

A consequence of the vastly improved processing and communication of information, together with the ability to readily integrate information on a range of specialist areas, is a much greater capability to meet the individual needs of customers. Mass solutions are no longer necessary in a world where individual preferences can be easily incorporated into ordering systems. Separate products are no longer necessary when systems exist to provide a fully integrated product to the customer.

If products are to be customised for users then there will need to be standardised user friendly descriptions of the attributes of a product to which a customer can react. Since regulatory requirements will differ from one country to another, and customer choices will differ where regulations do not exist, a standard prescribing a single universal level of strength is not acceptable. Furthermore it needs to be in terms that can be understood, not by engineers, but by purchasers of the product being designed, be it the whole building, or sub-assemblies such as facades or precast floor elements. Customisation of standards thus requires standardisation of the description of performance in a way that allows different levels of performance to be described in a simple user friendly manner, together with the more traditional technical standards against which the actual certification of the product can be made.

3.5.2 Trade Globalisation

Globalisation of trade implies an increasing focus on international standards at the expense of national standards, and an increasing demand for standards. Indeed the proliferation of standards, and particularly international standards, in recent years already reflects this impact. The International Organisation for Standardisation (ISO) provides a vehicle for the production of these international standards, and the infrastructure for an international system of product certification is being established. In order to be universally acceptable these standards must also be performance standards not prescriptive standards. This has been recognised by the World Trade Organisation (WTO) as discussed in 3.4.3.

A number of steps need to be taken if a fully integrated international system for trading of building products and services is to be established: (a) deregulating standards, (b) international certification system, (c) international standards for the description of performance.

(a) Deregulating standards.

Deregulating standards is an essential first step towards a fully integrated international system. Regulations are a statement of politically acceptable minimum requirements. Because of their political nature, they may vary from community to community, or from state to state, or from country to country. Hopefully with the passage of time and the development of a single global market place they will move closer together, so removing the barriers to trade that they often impose. But this will require political action and political will. Building standards on the other hand are essentially technical statements that standardise, generally in terms of quality or performance, but sometimes in terms of size or procedure, some activity in relation to building and construction. They can be national or international. Even if developed for one community, they should be quite capable of being used in other communities providing they meet that community's minimum regulatory requirements. If building and construction is to become an integral part of the international market place, this seems the most practical way for it to occur.

(b) International certification system.

There is an increasing consumer needs for assurance of both technical and manufacturing quality. Certification of products against recognised standards can be expected to be a major driving force, particularly if purchasing the product from a global market.

(c) International Standards for the description of performance

It follows therefore that the first step in the development of international standards should be the standardisation of the description and measurement of performance levels, which are universal in their application, and against which existing standards can be calibrated and regulatory requirements bench marked.

This would provide a clear path for acceptance of international standards by regulators in different countries, a standardised reference for the international insurance industry in dealing with a multiplicity of codes and standards, and a basis for an internationally recognised system of certification of building products in terms of technical and manufacturing quality.

(d) Performance-based Regulations

Such an approach is only possible if building regulations are developed in performance format. Some countries' building codes are already well advanced in this respect, enabling them to readily incorporate deregulated standards of the type described. However, a performance-based system can create favourable conditions for trade only if it can provide: (a) clear intents via quantified performance requirements, (b) common basis for comparing and assessing products. Most of current 'performance-based' systems fall well short of this ideal.

3.5.3 Risk-based Approach

The events of September 11, 2001 will have an impact on the way tall buildings are designed and constructed. Various investigations are being undertaken by the US National Institute of Standards and Technology (NIST), the International Council for Research and Innovation in Building and Construction (CIB) and the US International Code Council (ICC). Underlying these investigations are a number of risk related issues such as the level of risk considered to be tolerable to society, ranges of extreme events to be considered in design, role of regulations in mitigating risks, technological capabilities for assessing and mitigating extreme events etc.

Other lessons for building construction revolving around the concept of robustness including redundancy in structure and fire design, load combinations involving fire loads for critical members and connections and impact resistance of structures.

These investigations are still at an early stage and the outcomes in terms of changes to regulation and standards are still to come.

3.5.4 Environmental Concerns

The construction industry and the built environment are two key areas in sustainable development. For example, in Europe, buildings are responsible for more than 40% of total energy consumption and the construction sector generates about 40% of all man-made waste (CIB, 1999). Issues in sustainable construction that may affect structural design consideration include embodied energy (eg. material selection, reuse, recycle) and climate change (eg. changes in design environmental loadings, durability) etc.

Most of developments, in regulations and standards toward sustainable development, have so far focussed on operational energy efficiency and have little effects on structural design. However, sustainable development issues will, in time, assume increasing importance. In terms of building products, the issues are to reduce the amount of embodied energy, to lower emissions from products in use and to improve restorability and recyclability. Indoor air quality requirements may also restrict the use of certain materials in construction.

Greater attention also must be given to the effects of climate change including the effects of warmer climate, rise in sea level and increases in rainfall, flooding, wind speeds and storm surges. The number of extreme events and their severity will be greatly increased. A better understanding and quantification of these effects are needed before any changes to regulations and standards can be contemplated.

Structural engineering design in future will have to come to term with these issues and must operate as part of an overall strategy for sustainable development.

Reference

Edwards, C. (1904) The Hammurabi Code and the Sinaitic Legislation, with a complete translation of the great Babylonian inscription at Susa, Watts, London

ISO 6240 (1980) Performance standards in buildings - Part 1: Contents and presentation

ISO 6241 (1984) Performance standards in buildings - Part 2: Principles for their preparation and factors to be considered

ISO 7162 (1992) Performance standards in buildings - Contents and format of standards for evaluation of performance

ISO 9699 (1994) Performance standards in buildings - Checklist for briefing-Contents of brief for building design

ISO 2394 (1998) General principles on reliability for structures.

NBS (1972) Performance Concept in Buildings - NBS Special Publication 361 US Dept. of Commerce

Galambos, T.V., B. R. Ellingwood, J.G. MacGregor and C.A. Cornell (1982), "Probability based load criteria: assessment of current design practice," *J. Struct. Div. ASCE* 108(5):959-977.

Ellingwood, B. R., J.G. MacGregor, T.V. Galambos and C.A. Cornell (1982), "Probability based load criteria: load factaors and load combinations," *J. Struct. Div. ASCE* 108(5):978-997.

ECE (1996) ECE Compendium of model provisions for building regulations, United Nation Publication, Geneva

IRCC (1998) Guidelines for the introduction of performance-based building regulations, The inter-jurisdictional Regulatory Collaboration Committee

Vienna Agreement (1991) Agreement on technical cooperation between ISO and CEN - ISO/CEN - First revision 1995, Second revision 1998

ISO/IEC (1992) Directive Part 2 - Methodology for the development of International Standards

ISO/IEC Guide 2:1991 - General terms and their definitions concerning standardization and related activities

WTO (1994) The WTO Agreement on Technical Barriers to Trade - World Trade Organisation, Geneva, Switzerland

Okada (2002) Basis of Design across Civil Engineering and Architecture (Japan) - Alignment of Codes and Standards in Building Construction - 7thWorkshop, Bali, Indonesia, March

APEC (1996-2002) Proceedings of APEC Workshops on International Alignment of Standards in Building Construction Melbourne (1996), Singapore (1998), Shenzhen (1999), Melbourne (2000), Hanoi (2001), Bali (2002).

EN 1990 (2001) - Eurocode: Basis of structural design - August

AISC (1986) Load and resistance factor Design Specification for Structural Steel Buildings - American Institute of Steel Construction Inc. Chicago, Illinois, September

Gulvanessian and Driscoll (2001) Eurocodes - The new environment for structural design - Proceedings of the Institution of Civil Engineers Vol.144, November

NRC(1995) Objective-Based Codes in Canada - National Research Council, Ottawa, Canada

ABCB(1996) Building Code of Australia - Australian Building Codes Board

BIA(1993) New Zealand Building Code - Building Industry Authority of New Zealand

ICC(2000) Performance Building Code for Buildings and Facilities - International Code Council USA

HMSO (1994) The Building Regulation 1991- Approved Document A - Structure Dept. of the Environment and the Welsh Office - 1992 Edition Fourth Impression with Amendments 1994

CIB (1999) Agenda 21 on sustainable construction - CIB Report Publication 237

Chapter 4

Load Modeling

By:
Jun Kanda, University of Tokyo

4.1. Introduction

Presented herein is a discussion of load modeling for the structural design of tall buildings. The appropriate evaluation of load is an essential first step for the structural calculations to achieve the safety requirements. Safety requirements are often the major objective for ultimate limit state criteria in the design procedure. We are also concerned about the maximum of loads for a structure. The maximum load over the service lifetime is a basic variable to be considered in the design procedure, although the annual maximum load is another basic model and sometimes could be an alternative to the lifetime maximum load when the annual probability of failure is considered. The probabilistic nature of load which varies in time and in space is quite complicated but predominant cases to be considered are rather limited. Fundamental issues are introduced in this chapter. Detailed procedures for the probabilistic treatment of load models are available (Hart 1982, Wen 1990).

This chapter focuses the characteristics of major loads for tall buildings to be applied in the reliability-based design and as the reliability analysis. The characteristics of loads are different due to the physical mechanism. The Gravity is acting all the time on the earth and causes the vertical force on the mass of any object. The dead load (self weight), live load and snow load are all caused by the gravity. The wind causes the pressure distribution on the surface of buildings and the dynamic force results in an action on the structure. The earthquake load is caused by the ground motion caused by the earthquake. Mathematical models are used to simplify these loads of physical nature. Significance of parameters is discussed and probabilistic natures of load intensities are described.

4.2. Loads, Load Effects and Design Loads

The load evaluation plays an essential role in structural calculations. The reliability analysis is performed based on probabilistic load models. Loads are random variables and act on structures. Load effects are responses of structures and include the uncertainty of structural parameters such as the dynamic characteristics. The design load is a deterministic value to be used in the design procedure to check if the structure satisfies the requirements. The Reliability-based design provides a basic procedure for determining the design value for load by considering the variability of parameters and the target reliability as described in Chapter 6.

Load effects are to be compared with the resistance, strength, deformation or capacity of the structure through design formula. Various loads act simultaneously and load combinations are considered for appropriate design situations. Dead load and live load act the entire time and wind load or earthquake load act in a very short period of time. Therefore the wind or earthquake is combined with the dead and live loads but the wind and earthquake is not usually combined. A convenient combination rule called Turkstra's rule (Turkstra and Madsen 1980) is often used in design procedures. The maximum of combined load effects are expected to occur when one of individual load attains its maximum. Load combination cases are considered for cases when each load attains its maximum, then other loads are combined with usual values. Turkstra's rule provides a slight underestimation for combined load with prescribed probability, but is regarded as being reasonably practical.

Load effects, Q consist of the load intensity, X, and other parameters, C_i, often in a form of multiplication such as;

$$Q = X \cdot C_1 \cdot C_2 \cdots C_N$$

(1)

All these parameters in eq. (1) are considered as random variables.

The load intensity, X, is assumed to follow a probability model and is often estimated according to the statistics. Very rare events do not provide sufficient amount of data and physical models are used to compensate for statistical error or modeling uncertainty. But the fundamental variability of the load usually cannot be controlled by engineering processes since it occurs naturally.

Other parameters include environmental influence factors and transforming factors from the load to load effect. They are rather dependent on the engineering development and technical information. Their uncertainties can often be reduced by additional efforts and investigations. C_i is often treated as independent of X for the simplicity, but in reality they are dependent of X and in such cases the function becomes non-linear to X.

For the convenience of design formula, the design load Q_d is expressed as the load factor, γ, times the representative value Q_r which is obtained from a characteristic value of X, which is often defined as a value corresponding to the return period of 50 or 100 years for annual maximum events, and the mean of C_i. This is referred to as the load factor format or partial safety factor format (ISO2394, 1998). The load factor is obtained with a specified degree of safety according to the reliability analysis described in Chapter 5.

$$Q_d = \gamma Q_r \tag{2}$$

When the single load factor is used, as in eq. (2), γ should reflect the uncertainties of all parameters, i.e. X and C_i in eq. (1). But when the variability of C_i is not clearly specified or statistical information is not sufficient for C_i, γ is obtained by only considering the variability of X. The uncertainty of C_i can be reduced when sufficient reliable data are obtained. For example, wind tunnel experiments can be used to obtain the wind pressure or force coefficient and soil investigations are applied for reliable estimation of soil amplification factor for such purposes. These efforts could reduce the uncertainty of load effects and lead to the reduction of γ with the constant target safety.

Codes and regulations often specify γ and Q_r or directly Q_d. The return period or the probability of exceed for a reference period can be determined for Q_r in order to provide a unity load factor. The load factor could still reflect the reliability measure for the return period or the probability but the variability of other parameters, Ci is not explicitly included for the formulation of the load factor. Engineers have to keep in mind the quantitative evaluation of the variability and the uncertainty of major parameters involved in the load formulae.

Detailed formulation of design load with respect to the reliability index is referred to Chapter 6. Owing to the development of the Advanced First Order Second Moment (AFOSM) Reliability, the relation between the design values of load or resistance and the target reliability or the safety measure becomes clearly visible. Cost-benefit aspects are introduced to provide a rational assessment to the optimum safety or reliability of tall buildings.

Typical loads such as live load, wind load and earthquake load are discussed in this chapter.

4.3. Maximum Load

In the mathematical model of load, load intensity X is assumed to follow a probability model. Some specific probability models are mathematically derived from only the condition of being maximum (Gumbel 1957). The appropriateness of adoption of these distributions has to be discussed; nevertheless it is convenient to extrapolate the tail of the probability distribution for the future from a finite amount of past data.

Extreme value distributions i.e., the Gumbel, Frechet and Weibul distributions are three theoretical distributions that are commonly applied to model the load intensity parameters such as the annual maximum snow depth, the annual maximum wind speed or the annual maximum peak ground acceleration.

The Gumbel distribution is often applied to the maximum wind speed and the maximum snow depth, or water-equivalent snow load on the ground.

$$F_G = e^{-e^{-a(x-b)}}$$

(3)

where a and b are parameters and relations with the mean of x, μ and the standard deviation σ are expressed as;

$$a \cong 1.28 \frac{1}{\sigma}, \quad b \cong \mu - 0.45\sigma$$

(2a)

The Frechet distribution with lower bound value of 0 has been used for the maximum peak ground acceleration (PGA) or peak velocity.

$$F_F = e^{-\left(\frac{c}{x}\right)^k}$$

(4)

where parameters c and k are related to the mean μ and the standard deviation σ as;

$$\mu = c\Gamma\left(1-\frac{1}{k}\right) \quad \text{for} \quad k>1 \ , \qquad \sigma = c\sqrt{\Gamma\left(1-\frac{2}{k}\right)-\Gamma^2\left(1-\frac{1}{k}\right)} \quad \text{for} \quad k>2$$

(5)

When the data period is limited, such as 100 years, the upper bound tendency does not appear for the maximum ground motion intensity. For data in a much longer period, such as 1000 years, the saturation tendency appears. In other words, a model with the upper bound for the maximum value seems to be more realistic.

An empirical distribution with both upper and lower bound proposed by Kanda (Kanda 1994) is sometimes convenient to model extremes with large variabilities as a saturation tendency is often observed even for very rare natural events and the Frechet distribution cannot represent realistic tail characteristics.

$$F_K = e^{-\left(\frac{w-k}{ux}\right)^k}$$

(6)

where w is the upper bound value and 0 is used as the lower bound value as commonly used for the Frechet distribution.

These extreme value distributions are formulated as asymptotic distributions for the maximum of an infinite number of events and may not necessarily fit the data of natural phenomena particularly well. Nevertheless it is convenient to apply such forms to represent the maximum load intensity to capture the tail characteristics of the distributions which is extremely important to understanding the nature of loads. Furthermore the n-th power of the cumulative distribution function of annual maximum, $F_{ann}(x)$ is the cumulative distribution function of n-year maximum, $F_n(x)$, when the assumption is made that the annual maxima are identically distributed and statistically independent variables.

$$F_n(x) = \{F_{ann}(x)\}^n$$

(7)

The same formula for an arbitrary, period can be used for each of these extreme value distributions with different parameter values, as simply examined when one of eqs.(3), (4) and (6) is substituted in eq.(7). If sufficient data are available and only the interpolation is needed, the application of extreme value distribution may not be necessary. However the data inevitably are insufficient and therefore the extrapolation is inevitable based on the extreme value distribution assumption to evaluate the load intensity with a very low probability of exceedance.

The period 50 years is used as the reference period for the reliability-based design as it could be regarded as a representative period for a lifetime of a building structure and a period which can be discussed within a single person's experience. Eq.(7) is convenient for engineers to find out the relation of maximum value of load intensity for different periods.

4.4. Live Load

Live loads are often specifically given by building occupancy in codes. Statistics are informative to practitioners to understand the variability of live load distribution. According to some live load statistics, the gamma distribution models the variation of live load intensity with respect to the floor area.

$$f(x) = \frac{\nu (\nu x)^{k-1}}{\Gamma(k)} e^{-\nu x}$$

(8)

where the mean $\mu = k/\nu$, and the standard deviation $\sigma = \sqrt{k}/\nu$. The gamma distribution with $k = 1$ corresponds to the exponential distribution and it approaches the Gauss distribution as k becomes greater.

The live load formula is often expressed as,

$$L = w_o C_E C_{R1} C_{R2}$$

(9)

where w_o is a load intensity in terms of the average live load for the influence area, C_E is a conversion factor for the Equivalent Uniformly Distributed Load (E.U.D.L.), C_{R1} is a reduction factor with the influence area, and C_{R2} is a reduction factor for the number of floors to be considered.

Because every possible setting of furniture and persons can not be treated in the design calculation,,, the E.U.D.L., which is the uniformly distributed load producing the same load effects as individual setting of load patterns, is used. The statistics are based on arbitrary point in time information. Then the extra-ordinary live load has to be simulated according to a scenario which realistically causes a high concentration of loads. The lifetime maximum live load is modeled according to such scenarios with their frequencies. Furniture concentrations during removals and people's clustering at emergency situations can be considered as typical scenarios for extra-ordinary live loads as studied by Corotis and Tsay (1983) where the maximum load is modeled as the summation of sustained load and the extraordinary load or as studied by, Kanda and Kinoshita (1985) and Kanda and Yamamura (1989) where the maximum load is modeled as the extraordinary load including sustained load with different concentrations. The 50 year mean recurrence interval value of maximum live load is approximately equal to the nominal live load often found in national standards.

The time variation of live load can be modeled by a combination of rectangular pulse representing mostly furniture load and intermittent impulse representing furniture or people's concentration. A schematic diagram is shown in Figure 1. An average period of occupancy change is 7 to 10 years for office buildings, so the lifetime maximum cumulative probability distribution can be obtained by the 5th to 7th power of that of arbitrary point in time maximum.

Since the distribution nature of live load is not perfectly correlated, the variability of live load for a wider area tends to decrease. The same situation applies to the column load carrying multiple floors. C_{R1} and C_{R2} are formulated according to the statistics by considering the correlation effects. The probabilistic nature is simply explained in terms of spatial variability for rooms and buildings (Ellingwood and Culver, 1977).

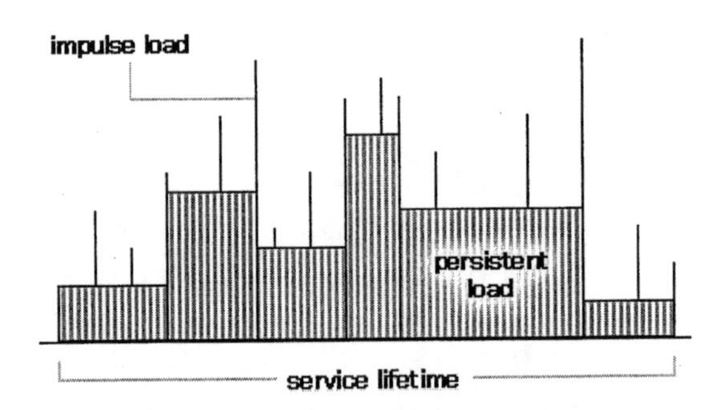

Figure 1 A schematic diagram of live load variation in time

4.5. Wind Load

The natural wind for structural design is fairly well described. The wind is modeled as an air stream distributed vertically with fluctuations in a turbulent boundary layer. This is an outcome of active wind engineering research since 1960's when Dr. Davenport

(1961), developed the gust response model from turbulence spectra model to describe the equivalent static wind load. The wind response due to the turbulence is called buffeting and the gust response factor is commonly adopted in wind design codes in most countries, particularly for tall buildings. Comprehensive explanations and derivations of formulae together with technical information on wind effects are available in textbooks such by Simiu and Scanlan, 1996.

A typical wind load formula can be expressed as;

$$W = \frac{1}{2}\rho(U_o E_H)^2 C_f G_f A \qquad (10)$$

where ρ is the air mass density, U_o is the basic wind speed, which is defined as a maximum wind speed averaged in ten minutes at the height of 10m above ground on a representative roughness condition, E_H is the vertical profile coefficient for wind speed at the height of building, which is given according to the wind profile model. C_f is the wind force coefficient basically obtained by a wind tunnel test but given in a code for typical building shapes, G_f is the gust response factor and often specified in a code, and A is the projected area.

The wind load formula for the cladding can be similar to eq(10) but with C_f as the difference of wind pressure coefficient on external cladding and interior pressure, and with G_f as the gust factor considering the cladding area. Even when the design earthquake load is the dominant horizontal load, cladding loads are governed by the wind in most cases.

The global wind force occurs as summation of wind pressure distributed over the surface. The wind pressure on a windward face increases with height similarly to the vertical profile of the approaching wind speed and on the leeward face it is fairly uniform in the wake region. Such characteristics are reflected in specified values for C_f in most building codes. For cladding, the wind force is obtained as the difference of wind pressure on the external face and internal face. C_f and G_f for the cladding are different from those for the force acting on the whole structure, but the similar formula as eq.(10) can be used.

The wind speed statistics are available from meteorological stations all over the world. National or state wind codes often provide a wind hazard map in terms of basic wind speed which is defined such as a ten-minute mean wind speed over a reference open terrain corresponding to a specified return period in years. In order to estimate the characteristic value with reasonable accuracy, sufficient data corresponding to the return period is required. Therefore in some cases a very rare event such as a typhoon or hurricane has to be examined based on climatological and physical models. One example of a simplified procedure for estimating hurricane winds is proposed by Sanchez-Sesma, Aguirre and Sen (1988).

The Gumbel distribution has been commonly used to model the annual maximum wind speed. Directional characteristics are also an important issue for the safety criteria for wind response. Some probabilistic models for directional wind characteristics are available (Itoi, T. and Kanda, J. 2002). The wind statistics sometimes do not appropriately provide the directional information. Wind measurements at a construction site may provide additional data for wind statistics particularly where local topography effects are expected. The period for measurements at a site is usually very limited and so should not be paid too much weight for extreme value estimation but the data is still useful for a wind speed model for serviceability criteria.

The spectral characteristics are given in a form of the power spectral density and the coherence function. Minor differences may be pointed out among models proposed by various researchers for the wind turbulence but results, in terms of the gust response factor, do not differ much from an engineering point of view. This means that the buffeting vibration is rather well described by the analytical manner. On the other hand, the across-wind vibration is a little more complicated as the dynamic wind load is more sensitive to the shape of structure. A predominant frequency for the dynamic across-wind force can be observed as clearly as the vortex shedding excitation in a smooth flow. Even in a turbulent flow a predominant frequency of the dynamic across-wind force can be observed. Then a peak value appears in the curve of the across-wind response with respect to the wind speed.

When the building becomes lighter and the fundamental period of building becomes longer, some of sections produce negative aerodynamic damping which leads to the self-exciting vibration, such as the galloping for the flexural mode and the flutter for the torsional mode. In most cases the galloping or flutter phenomena occur in a higher wind speed range than the vortex shedding excitation. The Scruton number is a good indicator for the sensitivity of structures to the across-wind vibration and sometimes used as a measure for the necessity of dynamic wind tunnel experiment. The Scruton number S_c is given by;

$$S_c = h\rho_s / \rho \qquad (11)$$

where h is the damping ratio of a structure and ρ_s is the averaged density of structure.

Some empirical formulae are available for across-wind vibration of tall buildings with a simple shape as a form of spectral density of dynamic across-wind force (A.I.J. 1996). Although the along-wind buffeting response, as formulated in eq. (10), is often a major wind load effect, the across-wind response has to be investigated for tall buildings in particular when the serviceability for vibration perception is considered.

Wind tunnel experiments are generally recommended for tall buildings especially for cladding design as the local wind pressure distribution is sensitive to the shape of building. A five component balance apparatus can be used to measure dynamic wind fluctuations. When negative aerodynamic damping due to the interaction effect is expected, a wind tunnel experiment for aero-elastic model is recommended. The quality of wind tunnel experiments is carefully examined for the similarity of simulated turbulent wind to the natural wind and the accuracy of pressure and force measurements. Manuals are often available for high standard experiments, e.g. ASCE Manual (1987).

Damping devices are often installed in tall buildings to suppress the wind response mostly to meet the serviceability criteria or improve habitability. The increase of damping from the order of 1% to 5% could reduce the response by approximately one half and many successful results have been reported. The damping value is dependent on the response level. Design damping ratios for safety evaluation are recommended based on Japanese building database as 3.0% to 1.2% for reinforced concrete buildings with height from 30m to 100m respectively and as 2.0% to 1% for steel buildings with height from 30m to 200m respectively (Tamura, Suda and Sasaki, 2000).

4.6. Earthquake Load

Earthquake load is specified in a rather hypothetical manner as it is not a direct load such as the gravity or the wind force, but is treated as a load for the structural calculation to provide load effects as resulted by the ground motion caused by earthquakes.

An earthquake hazard map is commonly used to specify a characteristic value for the seismic load intensity in many countries. However, it is not well recognized among practitioners to treat the seismic load as a random variable. The prediction of earthquake occurrence includes a high uncertainty; the fault model and the attenuation model also have always high uncertainties. The soil amplification also has to be modeled to predict an input motion to a tall building. When these uncertainties are quantified, seismic load effects are described as random variables to be used in a reliability-based design. Advances in seismology have to be reflected in the earthquake hazard analysis. Estimation of the occurrence probability of earthquake for a specific active fault can improve the hazard map.

Recently, damping devices have been introduced in many tall buildings in Japan to suppress the response motion, i.e., to reduce the load effects. Viscous liquid dampers and in-elastic deformation energy absorption dampers both have been developed.

Equivalent static earthquake load for i-th the level in terms of lateral force can be expressed as in ISO 3010 (2001),

$$F_i = k_E k_D k_R k_{F,i} \sum_{j=i}^{n} F_{G,j} \tag{12}$$

where k_E represents the earthquake load intensity with a specified probability of exceedance or return period, k_D is structural factor considering the ductility or other structural characteristics, k_R is the response spectral factor considering the soil conditions and the structural damping factor, $k_{F,i}$ is the force distribution factor for the i-th level and $F_{G,j}$ is the gravity load of the j-th level.

The load intensity in a hazard map provides either the peak ground acceleration (PGA) or peak ground velocity (PGV) for a reference hard soil condition. k_E in equation (12) is given in terms of PGA normalized by the gravity acceleration. When the spectral characteristics are specified in the equivalent static load, only the load intensity k_E can scale the load effect. However the spectral characteristics of motion influence the ductility parameter and the duration of motion also influences the cumulative energy absorption or the in-elastic deformation. Therefore a dynamic analysis in the in-elastic range is usually conducted for tall buildings for the safety evaluation. Some of recorded motions such as El Centro 1940 or Taft 1957 have been used with its scaled PGV values, but the soil amplification characteristics are dependent on the local site and recorded motions cannot represent the spectral characteristics at the construction site, then one or some simulated earthquake time histories with appropriate spectral characteristics are preferred.

The restoring force characteristics are obtained for each level by the so-called push over analysis which is a static in-elastic analysis with a step-by-step increasing lateral load such as specified by equation (12). In order to provide the design input motion, its spectral characteristics have to be hazard consistent. In other, words the hazard level in terms of probability has to be consistent to the design input motion. There are two procedures for this consistency. In the first, the design load intensity is determined e.g. the representative value is multiplied by the load factor corresponding to the target reliability then the time history motion is simulated with a specified response spectrum considering the soil conditions. In the second, hazard consistent magnitude and distance for a hypothetical fault is determined corresponding to the target reliability then the design time history is simulated for the fault considering the attenuation path and also soil conditions of the site. When particular active faults are considered in the structural design, their occurrence probability and the uncertainty of fault models have to be examined and compared with the hazard consistent design input motion in order to provide a reliability measure. The nature of load intensities for wind load and earthquake load can be compared with emphasis on their probabilistic characteristics a simplified comparison is shown in

Item	Wind	earthquake
Basic intensity	gradient wind	bedrock motion
Typical probability distribution	Gumbel	Frechet
c.o.v. for 50 year maximum	10%-20%	40%-100%
transmitting medium	Air	soil
cause of dynamic action	boundary layer turbulence	surface soil amplification
typical spectrum form	Karman spectrum	Kanai-Tajimi spectrum[1)]
predominant frequency	0.02Hz-0.05Hz	0.5Hz-2Hz
duration time	10min-1hour	10sec-2min

Table 1 Simplified comparison of wind speeds and earthquake ground motions from a view point of stochastic nature (Kanda and Nishijima 2002)

4.7. Other Loads

The structural design should pay attentions to all possible loads influencing the structure. Dead load, snow load, temperature load, soil and water pressure are all important loads. Fire load is also an important load as shown by the September 11th events.

When the mass density of every material used for the building is carefully examined and the size of every component and member is carefully measured, the uncertainty of the dead load will be reduced into a sufficiently small amount in comparison with other loads. However the accuracy of the physical conditions assumed in structural calculations and minor changes of material or member during construction could increase cause the variability of dead load.

The snow load is sometimes a very significant gravity load in heavy snow regions. The safety against the snow load is very critical for a light roof. The temperature load must also be considered for a long span roof which is expected to have temperature differences due to the sunshine and the shade. The soil pressure and the underground water pressure are discussed in chapter 8.

Structural action due to fire is actually an imposed deformation, analyzed as a response to a specified fire exposure curve describing the relation between temperature and time. The fire exposure is a function of the weight of combustibles in the compartment (related to the building contents), the ventilation of the compartment, and the bounding surfaces of the compartment. Prolonged exposure to fire causes the strength and stiffness of structural elements to decrease leading to very large deformations and eventually the onset of structural collapse.

Various situations are considered for the fire safety of tall buildings. (CTBUH 1995) An unexpected extraordinary event such as the airplane crash on the September 11, 2001 can not be dealt with an ordinary structural design. Nevertheless the robustness of structural system has to be carefully considered particularly for a building with a huge capacity of people in order to minimize the failure consequences.

4.8. Target Safety Criteria for Loads

The life-cycle cost assessment is now of concern by not only engineers but clients and architects. The expected failure loss is a part of the life-cycle cost and the target reliability can be optimized by minimizing the life-cycle cost including the initial cost, the expected failure cost and the maintenance cost. The failure cost model for a tall building has to be examined to provide a rational explanation for the target safety criteria.

The life-cycle reliability based design is described in Chapter 6. As the design load is directly controlled by the target reliability, some examinations on the life-cycle cost from a view point of load are demonstrated.

A simple expected total cost can be written as,

$$C_T = C_I + P_f C_F \tag{13}$$

where C_T is the total cost, C_I is the initial cost, P_f is the probability of failure and C_F is the failure cost. For individual buildings, different maintenance costs will be considered and many failure states have to also be considered for a more realistic life cycle cost evaluation. But equation (13) provides a basic nature of the balance between the cost and benefit where the initial cost is the initial cost controls the safety of structure and the benefit is the reduction of the probability of failure. In many cases the maintenance cost can be independent from the target reliability or the design load.

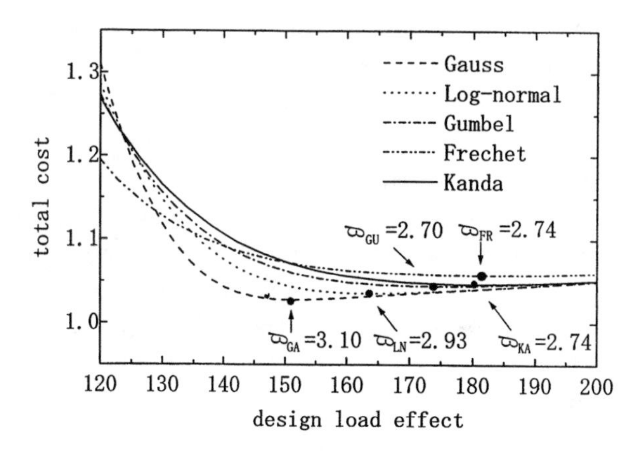

Figure 2 Total expected cost v.s. design load effect with different probability models

For a case of a single load effect and resistance, both of which are assumed as log-normal variables, a closed form solution is available (Kanda and Ellingwood, 1991). The normalized failure cost factor, the cost-up ratio and the coefficient of variation of load effect are equally important. An example is shown in figure 2 for the total cost with the design load effect where five different probability load models are used with typical parameters for ordinary structures (Kanda 2000). The mean of the maximum load effect is 100 and the c.o.v. is 30% and 300 was used as the upper bound limit for Kanda's distribution of equation (6). The general tendency of total cost curves against the load effect shows a sudden decrease and after the minimum values a gradual increase along the initial cost line with the design load. The Gauss distribution deduces the lowest design load with the highest reliability and the Frechet distribution leads to the highest design load with the lowest reliability.

The optimum reliability index and the corresponding design load are sensitive to the load model and the normalized failure cost which is a rather subjective parameter and the value itself has to be carefully utilized. Nevertheless, individual buildings have different environments and demands and such parameters as the failure cost, the cost-up ratio and load models have to be examined and discussed thoroughly at the beginning stage of structural design (Kanda and Shah, 1997).

4.9. Conclusions

Major loads for tall buildings are discussed in this chapter with emphasis on their probabilistic nature. General treatments in structural engineering for the live load, wind load and earthquake load are reviewed. Although practically sufficient data for the probability model of load intensity are accumulated and utilized for reliability-based design, there still exist many variables whose probability models are not sufficiently supported by statistical data. Engineers have the responsibility for the client and the society to determine the design load consistent to the appropriate safety degree. Loads are clearly specified in codes, standards or regulations, but regulations only provide the minimum requirements which often are not appropriate for the most advanced engineering achievements. Following the regulation is not sufficient and a continuous effort to stay current with advances in engineering is required.

References

Archit. Inst. Japan 1996, AIJ Recommendation for Loads on Buildings, Architectural Institute of Japan.

ASCE 1987, Wind Tunnel Model Studies of Buildings and Structures, ASCE Manuals and Reports on Engineering Practice No.67.

Corotis, R.B. and Tsay, Wen-Yang 1983, Probabilistic Load Duration Model for Live Loads, J. Struc. Eng., ASCE, 109, 859-874.

CTBUH 1995, Fire Safety in Tall Buildings, Council of Tall Buildings and Urban Habitat Committee 8A.

Davenport, A.G. 1961, The Application of Statistical Concepts to Wind Loading of Structures, Proc. I.C.E. 19, pp.449-472.

Ellingwood, B. and Culver, C. 1977, Analysis of Live Loads in Office Buildings, Proc. ASCE, J. Struc. Div., 103, 1551-1560.

Gumbel, 1958 Statistics of Extremes, Columbia Univ. Press

Hart, G. C. 1982, Uncertainty Analysis, Loads, and Safety in Structural Engineering, Prentice Hall.

I.S.O., 1998, ISO2394 General Principles of Reliability of Structures, Third edition.

I.S.O., 2001, ISO3010 Basis for Design of Structures – Seismic Actions on Structures.

Itoi, T. and Kanda, 2002, J. Comparison of Correlated Gumbel Probability Models for Directional Maximum Wind Speeds, J. Wind Engineering and Industrial Aerodynamics, 90, 1631-1644.

Kanda, J. 1994, Application of an Empirical Extreme Value Distribution to Load Models", *Journal of Research of the National Institute of Standards and Technology*, vol. 99, No. 4, 413-420

Kanda, J. 2000, Towards Rational Framework for Wind Loads on Buildings, Proc. Intern. Sympo. Wind and Structures, Cheju, Techno Press, 131-138.

Kanda, J. and Ellingwood, B. 1991, Formulation of Load Factors based on Optimum Reliability, Structural Safety, 9, 197-210.

Kanda, J. and Kinoshita, K. 1985, A Probabilistic Model for Live Load Extremes in Office Buildings, Proc. 4th Intern. Conf. Struc. Safety and Reliability, Kobe, II, 287-296

Kanda, J. and Nishijima, K. 2002, Wind Loads and Earthquake Ground Motions as Stochastic Processes, 1st ASRANet International Colloquium, Glasgow

Kanda, J. and Yamamura, K. 1989, Extraordinary Live Load Model in Retail Premises, Proc. 5th Intern. Conf. Struc. Safety and Reliability, San Francisco, III, 1799-1806.

Kanda, J. and Shah, H. 1997, Engineering Role in Failure Cost Evaluation for Buildings, Structural Safety, 19, 79-90.

Simiu, E. and Scanlan, R.H. 1996, Wind Effects of Structures, Third ed. John Wiley and Sons.

Sanchez-Sesma, J.J., Aguire, J.J. Sen, M. 1988, Simple Modeling Procedure for Estimation of Cyclonic Wind Speeds, J. Struc. Eng. 114, 352-270.

Tamura, Y., Suda, K. and Sasaki, A. 2000, Damping in Buildings for Wind Resistant Design, , Proc. Intern. Sympo. Wind and Structures, Cheju, Techno Press, 115-129.

Turkstra, C.J. and Madsen, H.O. Load Combination for Codified Structural Design, J. Struc. Div. Proc. ASCE, 106, 2527-2543.

Wen, Y.K. 1990, Structural Load Modeling and Combination for Performance and Safety Evaluation, Elsevier.

Chapter 5

Reliability Analysis

By:
R.E. Melchers, The University of Newcastle, Australia

5.1 Introduction

The essential concepts for the estimation of the reliability of a structure against adverse environmental effects can be illustrated very simply. This will be done in the section to follow. We will then refine the concept to account for multiple loads and for different types of loads. This will lead, with some mathematics, to the concept of a first exceedence event. For certain types of load processes and under certain conditions this problem can be solved but for most practical cases it has been found necessary to simplify the problem in various ways. These will be outlined briefly and one particular approach, the First Order Second Moment (FOSM) approach will be described in somewhat more detail, mainly because of its simplicity and its extensive use in practical problems. Remarks about the validity of the simplified approaches close the chapter.

5.2 Outline of Structural Reliability Problem

A structure such as a tall building will be subject to a variety of possible loading types, such as wind, earthquake and live loads as well as other loads such as due to snow or temperature effects. For the discussion to follow it will be sufficient to concentrate on wind and live loading. The other loads can be added easily once the basic ideas are understood.

Wind loading acting on a tall building can be visualized mainly as a fluctuating lateral force as might be represented in Figure 1. If the structural system strength against such loading is known exactly, it follows that the capacity of the structure is exceeded whenever the loading increases to a point on or above the capacity. This point is shown in Figure 1. Of particular interest for structural engineers is the time t_1 to the first occurrence of such an event (the 'first exceedence event'). This time should be long for safe structures. It can be estimated readily if the exact trace of windloading is known. Since the wind force at any point in time cannot be known except statistically, wind loading is usually modelled as a random process in time. This means that the time which has to elapse before the wind force first exceeds a given level (the 'first exceedence time') will be a random variable also. How we can estimate it is a central matter for structural reliability theory (although it is seldom actually calculated).

To the left of Figure 1 are shown two probability density functions. The main one (the 'instantaneous' distribution) refers to all possible wind loads described by the wind loading random process. Note the point where the resistance realization cuts across it. There is a small (shaded) part of the probability density function above this cut-off. This is the probability that the load will be greater than this value (for example, of structural resistance). Obviously the shaded zone will be smaller for higher values of resistance, that is, the probability that the load will exceed the strength will be smaller. Of course, the time to first exceedence will be longer also. It is clear, then, that the resistance level is very important. We shall return to this directly but first detour to note a commonly used terminology associated with loads.

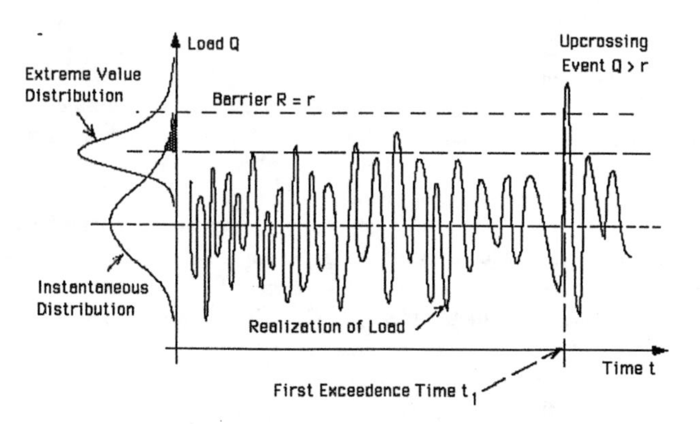

Figure 1 - Realization of windloading modelled as a continuous random process showing an exceedence of structural resistance and time to first exceedence.

Design wind (and other) loads are sometimes referred to as, say, the '1000-year' wind load. There is a value of wind load, say w_d, attached to this statement. This means that a wind load of intensity w_d or greater will occur, on average, once in a thousand years. Equivalently, such a load (or greater) has a probability of occurrence per year of about 0.001. In terms of Figure 1, w_d represents the resistance intercept and the shaded area is the probability of occurrence (not to scale). Sometimes only the maximum value of the load in a given time period (e.g. each year) is recorded. The corresponding probability distribution is then known as the 'extreme value' distribution (see Figure 1).

More precise discussions of these concepts are available in the literature. However, the important point is that statements such as 'w_d is the 1000-year wind load' do not tell us anything about loadings which are more likely to occur or which are less likely to occur (and therefore of greater magnitude). It is the latter which often are very significant for structural safety. And for that we need to know the probability of structural system failure, not just the exceedence probability.

The probability of structural system failure and the exceedence probability are not the same since, in practice, the actual resistance of the structure against wind (or other) load is not known precisely. The design strength has to be translated from the designer's specifications to the finished structure. It is well-known that this is not an exact process and that variations occur in material strengths and in workmanship. (We shall leave aside the question of gross errors and mistakes). Thus the exact location of the line $R = r$ in Figure 1 is not known precisely - all that is usually available is a probabilistic estimate.

The structural strength or resistance in reliability theory usually is modelled as a random variable. It follows that the actual strength is just one 'realization' of many possible design and construction outcomes - the line shown in Figure 1 must be considered in this way.

In estimating probability of failure we do not known which strength outcome applies to the particular structure in which we are interested. Therefore all possibilities must be considered even though some are more likely than others. The difference in likelihood of outcomes for structural resistance is expressed in the probability density function for the resistance. In some cases, such as for very simple systems where structural strength is directly measured by material strength, test information will provide a histogram from which the probability density function can be inferred (using standard methods for fitting distributions to data, such as the methods of moments, maximum likelihood, etc.). Figure 2 shows the situation schematically.

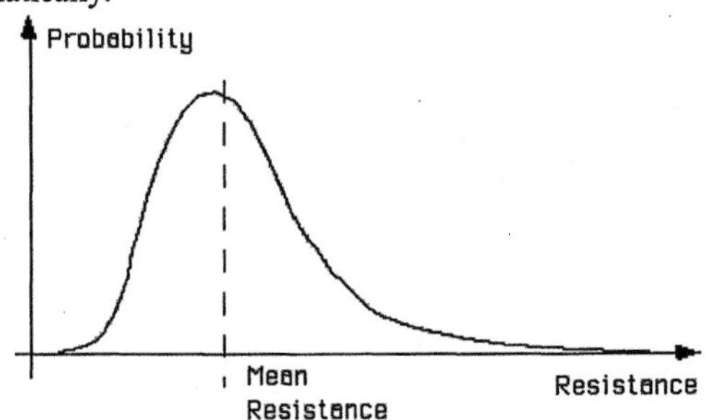

Figure 2 - Schematic probability density function of material strength.

More generally, uncertainty in material strength, structural and cross-sectional dimensions, workmanship, boundary conditions, etc. must be considered also. Thus if the strength of a cross-section consists of an uncertain material strength M and an uncertain cross-sectional area A, the strength S of the member (in tension, say) is given by

$$S = M.A \qquad (1)$$

The probability density function $f_S(s)$ of S can be determined from the corresponding probability density functions for M and A through (1). Unless these density functions are of simple form, this will require numerical integration. A particularly simple approach is to calculate just the first two moments of S - the mean μ_S and the variance, σ_S^2 respectively - using standard expressions, as follows:

$$\mu_S = \mu_M + \mu_A \tag{2}$$

$$V_S^2 \approx V_M^2 + V_A^2 \tag{3}$$

where $V = \sigma / \mu$ is the coefficient of variation.

So far we have considered only one load - wind loading. In real structures several loads may apply. Not all loads are modelled in the same way. Live load is often better modelled as a series of pulse loadings - such as when an office party or meeting causes a short-term high intensity loading. Again, when the loading increases to form a pulse it is possible for the strength of the structure to be exceeded. This might happen through beam or slab bending or shear or through column axial loading or through some combination of actions to be exceeded. Figure 3 shows this schematically.

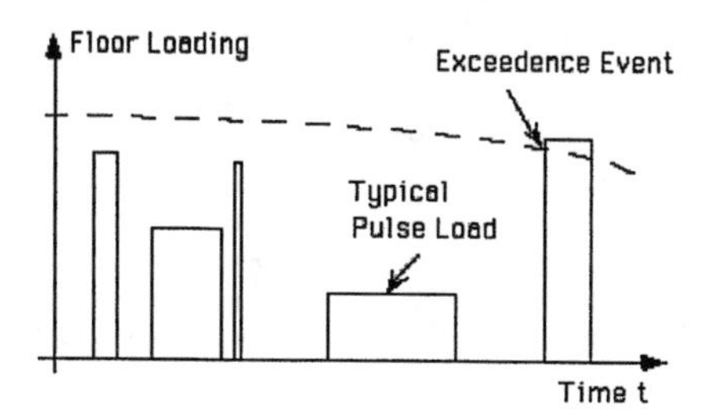

Figure 3 - Load modelled as a pulse process showing exceedence of structural resistance.

In practice both types of loading may act on the structure. Although in practice it is unlikely that both will be increasing at the same time, in principle, for a general theory, this possibility must be considered, as shown in Figure 4. The structural strength (resistance) in Figure 4 has been shown as some combination (an 'envelope') of that of Figures 1 and 3. This is because under combined structural action the capacity of the structure is generally different to that under individual loading systems acting alone - as examplified by column strength capacity under combined axial and bending action. Note the region under the envelope represent the so-called 'safe' domain and that outside it the 'failure' domain.

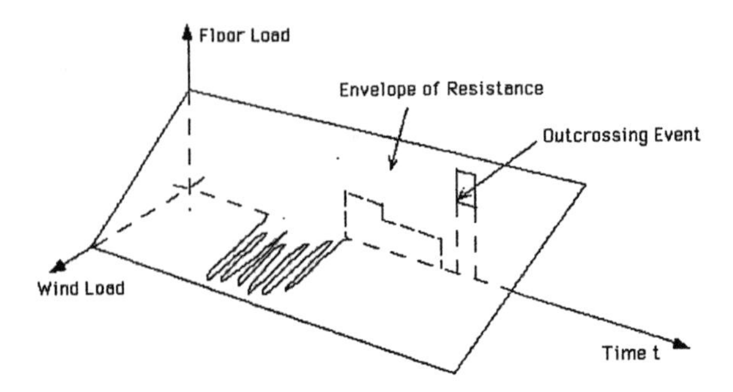

Figure 4 - Envelope of structural strength (resistance) showing a realization of the vector load process and an out-crossing by the floor loading component.

The first exceedence time is now the time when one of the load processes, or the combined action of the two processes, 'out-crosses' the envelope of structural capacity, that is, the combined process or either individually, cross from the safe domain to the failure domain. Figure 4 shows an outcrossing of the floor process.

The probability that the structure fails in a given time period $[0,t_L]$ (e.g. the 'design life') can be stated as the sum of probability that the structure will fail when it is first loaded, denoted $p_f(0,t_L)$, and the probability that it will fail subsequently (given that it has not failed earlier). This can be expressed as:

$$p_f(t) \approx p_f(0,t_L) + [1 - p_f(0,t_L)].[1 - e^{-\nu t}] \qquad (4)$$

where ν is the 'outcrossing rate'. The expression is approximate because the second [ ] term is based on the assumption that failure events are 'rare' and that such events therefore can be represented by a Poisson distribution. This then leads to the expression shown.

If we assume that the random load processes continue indefinitely and have a 'stationary' statistical nature (e.g. in the simplest case, their means and variances do not change with time), then the rate at which they would cross out of the safe domain (the 'out-crossing rate') can be estimated from:

$$\nu = \int_{\text{safe domain}} \mathrm{E}\left(\dot{X}_n \mid \mathbf{X} = \mathbf{x}\right)^{+} f_{\mathbf{X}}(\mathbf{x})\, d\mathbf{x} \qquad (5)$$

In this integral $\mathbf{X} = \mathbf{X}(t)$ represents the vector process and $(\,)^{+}$ denotes its positive component only, that is the component which crosses-out of the safe domain (the other component crosses back in and is of no interest). The term

$$E\left(\dot{X}_n \mid \mathbf{X} = \mathbf{x}\right) = \dot{x}_n = \mathbf{n}(t).\dot{\mathbf{X}}(t) > 0$$

represents the outward normal component of the vector process at the domain boundary and is there for mathematical completeness. Finally the term $f_{\mathbf{X}}(\mathbf{x})$ represent the probability that the process is actually at the boundary (since if the process is not there it cannot cross-out from the safe domain).

As noted, expression (5) is valid in the form given only for rare outcrossings such as might be associated with structural failure due to extremely rare high wind events. It would not be appropriate for, say, serviceability 'failure' as these would not, normally, be 'rare' events. The result can be extended to allow for gradual deterioration or enhancement of the structural strength with time with the result that $p_f(0,t_L)$ and ν become time dependent.

The central problem in structural reliability theory is how to evaluate expressions $p_f(0,t_L)$ and ν to estimate the expected life and hence the reliability of the structure. In the sections to follow we will discuss these matters.

In the foreseeable future, it is still unlikely that most structural designers will be concerned with the details of reliability calculations. Writers of modern design codes will have done this already and produced, from very many analyses, a much simpler approach to every-day design. This is the partial factor format familiar to structural engineers. The link between the theory concepts of this chapter and partial factor formats (of which there are several) will be explored briefly in Chapter 6. Alternatively, more detailed expositions may be consulted.

The situation is somewhat different for the evaluation of existing or ageing structures, particularly potentially hazardous systems or deteriorated ones. Most of these would not meet a review according to current design standards, yet many are still apparently satisfactory. For these, a more detailed and thorough safety and reliability review is appropriate. Although various subjective approaches to the safety and serviceability assessment of existing structures have been proposed, there is little doubt that thorough and defensible reviews will involve some degree of reliability theory. Detailed probability-based applications in conventional structural engineering have not yet been widespread but are not uncommon in the evaluation of structural aspects of, for example, nuclear power plants. Since older tall buildings are potentially hazardous, it is considered that application of reliability principles is appropriate for the safety assessments of these also.

5.3 Simplifications of Theory

The theory sketched above involves considerable computational work. It also demands information which may not be available. A simpler level of reliability estimation can be carried out at the expense of a small number of reasonable simplifications. These will now be described. The result will be a reliability estimation approach which does not deal explicitly with time as a variable.

5.3.1 Stationarity

The first step is to assume that the structural strength remains essentially constant with time, i.e. line $R = r$ in Figure 1 remains horizontal. This is reasonable provided the structural strength has not seriously been degraded by processes such as corrosion. It is also reasonable and conservative for concrete strength, since, although it usually increases with time, it is seldom the governing parameter in structural strength, except, in some cases, for shear.

The second step is to assume that the load processes are 'stationary', that is, their statistical properties do not change with time. This means that, typically, the mean of the load process is always the same, as is the variance (and higher moments, in general).

With these two assumptions, the outcrossing rate ν becomes constant with time. This means that the probability of structural failure for any time period (e.g. for each year) is constant. It also means that rather than considering the load process, it is sufficient to consider just its probability density function, that is, the pdf shown on the vertical axis in Figure 1. The loads can now be considered as random variables since one way of viewing a random process is as the limit of a sequence of random variables. However, some care is now required in how we treat several loads and how we combine them. We will discuss this later.

5.3.2 One extreme load only

To see how the loads must be defined, consider first a structural problem with just one load acting. Rather than considering it as a random process as in Figure 1, we could simply ask what the maximum value of the load is likely to be during the total life of the structure and with what probability this load will occur. This information can be obtained experimentally by recording the maximum load(s) on the structure each year and plotting a histogram. The probability density function we would infer from this is termed an 'extreme value' distribution (in this case for the maxima). It is shown in Figure 1 on the vertical axis.

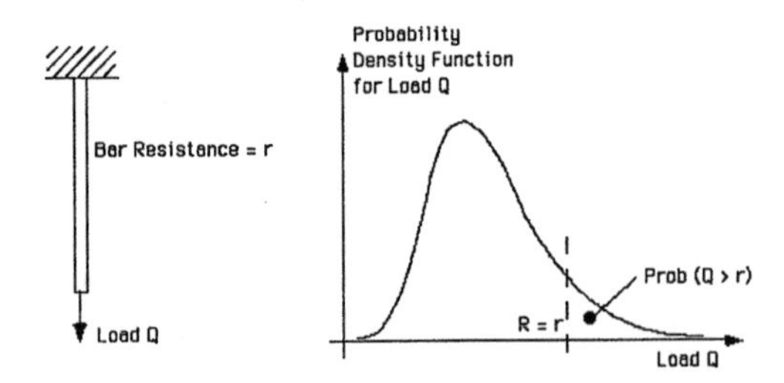

Figure 5 - (a) Bar in tension under applied load, (b) Probability density function of applied load.

The question of interest now becomes: what is the probability of structural failure in the event that the maximum load is applied to the structure? The failure event will occur only once and its probability of occurrence will depend on (i) the value of the maximum load (a random variable), (ii) the probability of occurrence of different possible values of candidate maximum loads and (iii) the actual strength of the structure as expressed by the probability density function for strength. The problem may be visualized in simple terms as shown in Figure 5(a). Here a structural member (a bar) is placed in tension by the application of a load. The actual value of the load is not known but it can be described by a probability density function $f_Q()$ with known mean and variance. (Note that typically $f_Q()$ will be an extreme value distribution).

For a reasonably well-designed structure, failure would not occur under low values of the load, but is increasingly likely as the applied load is of greater magnitude. If the strength of the bar is known exactly and is denoted by $R = r$ then the condition that the bar will fail is given by:

$$r < Q \qquad (6a)$$

or

$$Z = r - Q < 0 \qquad (6b)$$

where Z is known as the 'safety margin'. It follows that the probability that the bar will fail under the application of the load is equal to the probability that the load when applied is greater than $R = r$, or:

$$p_f = \text{Prob}(r < Q) = \text{Prob}(Z < 0) = \int_r^\infty f_Q(x)dx \qquad (7)$$

Realistically, the actual strength of the bar cannot be known precisely. Once the bar is constructed its strength can be estimated from tests on bars on similar materials and from its dimensions but some degree of uncertainty remains. Of course at design time, when decisions about the size of the bar and about acceptable safety might need to be made, there will be uncertainty also about the actual material strength and about construction techniques and their effectiveness, etc. In general, therefore, the strength of the bar must be expressed as a random variable also. Let this random variable be denoted R. The strength parameters such as mean μ_R and variance σ_R^2 and the probability density function $f_R()$ will vary with the extent of knowledge about the actual bar in the constructed structure - this need not be of concern at this time, except to note that past experience, manufacturers data and possibly in-situ testing may help to define them. Expression (7) can now be generalized to:

$$p_f = \text{Prob}(R < Q) = \text{Prob}(Z < 0) = \text{Prob}[G(\mathbf{X}) < 0] \qquad (8)$$

where $G()$ is known as the limit state (or performance) function and $\mathbf{X} = (R, Q)$ denotes the (random) vector of loads and resistances, in general. The expression $G(\mathbf{X}) < 0$ represents the condition that the bar will fail.

Let the probability distribution for the actual resistance be denoted $f_R(r)$. It then follows that (8) becomes:

$$p_f = \text{Prob}(R < Q) = \int_{-\infty}^{\infty} F_R(x) f_Q(x) dx \tag{9}$$

where $F_R(\,)$ is the cumulative distribution function for R. It is given by

$$F_R(r) = \text{Prob}(R < r) = \int_{-\infty}^{r} f_R(x) dx \tag{10}$$

Expression (9) is known as a 'convolution integral'. Loosely it can be interpreted as follows. Under the integral, the first term, given by (10), denotes the probability of failure given that the actual load has the value $Q = x$. The second term is the 'probability' that the load takes the value $Q = x$. This is then integrated over all possible values of x.

In general it is difficult to solve (9) in closed form. As will be seen below, an important exception is when R and Q are each represented with a normal distribution, or, more generally, are completely described only be their means and variances. Before doing so, however, comments will need to be made about (i) what happened to the initial failure probability $p_f(0,t_L)$ in (4), (ii) how structures with multiple loads can be treated and (iii) what happens when the structure becomes more complex than a simple bar.

Expressions (6-8) have ignored the initial failure probability $p_f(0,t_L)$ of (4). This is permissible for the way the problem has now been defined, with the maximum load applied (only once) at any time during the lifetime of the structure. The assumption implicit in this formulation is that the probability of failure is not affected by precisely when in the lifetime this event occurs.
Because the actual load process has now been represented as a random variable of the extreme load applied at any time during the life of the structure, the initial probability term of expression (4) has been subsumed into the random variable representation.

5.3.3 Structures with multiple loads

The above approach remains valid when there are several individual loads which are fully dependent on one another, that is, they are all representable by the one random variable. This may arise, approximately at least, for different points of wind loading on one wall. However, this situation does not apply when there is more than one loading system to be considered, for example, floor (e.g. crowd) loading and a wind loading. Typically these two would be considered as completely independent (although in some structures high wind loads might lead to low floor (crowd) loads).

It is not valid to represent each of a number of loads by an extreme value distribution, such as was done above. The reason for this is that it would not be expected that the occurrence of the maximum wind load acting on the structure would coincide exactly with the point in time when the maximum floor load would be applied, and similarly for other loads.

It is possible to develop the outcrossing approach to deal with this type of problem but for many purposes it is sufficient to apply a plausible and fairly accurate 'rule-of-thumb' known as Turkstra's rule. It states, simply, that the occurrence of the maximum of one load X_i should be associated with the 'average-point-in-time' values $\overline{X}_j$ of the other load(s). And out of these various combinations the maximum (or worst) combination will govern design. This may be stated as

$$\max X \approx \max\left(\max X_i + \sum_{j=1}^{n} \overline{X}_j\right) \quad j \neq 1;\ i = 1,\ldots,n$$

The 'average point in time' loads $\overline{X}_j$ are the loads which would be the loads expected to act on the structure if one were to measure them at any arbitrary time. For most loads, this means rather low load levels.

Turkstra's rule is adequate for many load combinations, such as where only one extreme load is likely to occur. But it is not satisfactory, for example, for the combination of snow load and high wind load when each of these are individually significant for a structure. In this case the load combination cannot be considered 'rare'. Similarly, if they are likely to be correlated, the combination of natural loadings is not well desribed by Turkstra's rule. An example is wind and wave loading for off-shore structures. However, for multi-story construction it would appear to be an adequate representation of load combinations since, say wind and earthquake loads are unlikely to be related. For this reason, it forms the basis for most modern limit states design approaches with a 'principal action - companion action' type of format.

5.4 Structural Systems

Usually structures consist of a collection of members or components. Mostly they cannot be modelled as shown in Figure 5. Except in special circumstances, it is not usually possible to predict which member or component will be critical to the reliability of the system. Potentially all members (and components) are critical since the precise interaction of local internal actions (consisting of the effect of one or more applied loads) and the member strength properties will depend on the random properties of the members and hence on their respective probability density functions.

In general, the reliability analysis of structural system is complex. Space precludes a detailed discussion. However, it is worth noting that many systems can be idealized as 'series' systems - in which failure of one member (or subsystem) is tantamount to failure of the system. This is the case, respectively, in brittle systems (for example a chain) and for the collapse modes of ideal-plastic systems - see Figure 6.

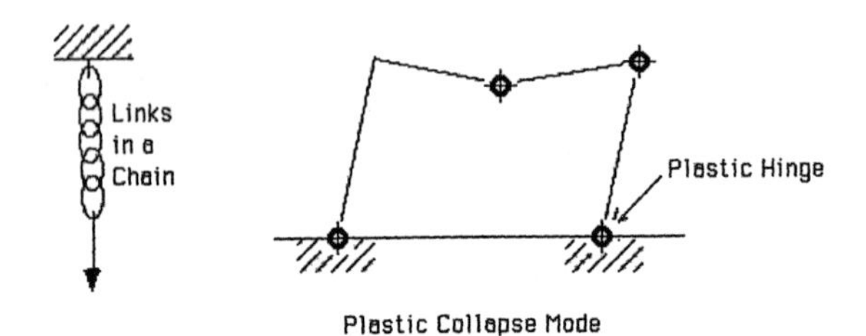

Figure 6 - Series systems

If the performance function $G_i(\mathbf{X}) < 0$ denotes failure of the ith member or in the ith collapse mode, series system failure is denoted by $\bigcup_i [G_i(\mathbf{X}) < 0]$.

If we now let $\mathbf{X}$ collect all the random variables in the reliability problem, the generalization of (9) and (10) becomes:

$$p_f = \int \cdots \int_{\bigcup_i G_i(\mathbf{X})<0} f_{\mathbf{X}}(\mathbf{x}) d\mathbf{x} \tag{11a}$$

where $f_{\mathbf{X}}(\,)$ is the joint density function of the random variables $\mathbf{X}$. The solution of (11) in general is not a simple matter. This is particularly the case if the limit state functions describe a complex failure domain and/or $f_{\mathbf{X}}(\,)$ is not of simple form. In addition, the dimension of $\mathbf{X}$ for realistic structural systems is often quite high.

If the region of integration in (11) is changed to the failure domain D_f in general, we have the general statement of the time-independent structural reliability problem:

$$p_f = \int \cdots \int_{D_f} f_{\mathbf{X}}(\mathbf{x}) d\mathbf{x} \tag{11b}$$

Evidently, the domain D_f will be specified by failure criteria (such as a maximum allowable stress, or a yield condition) and through a structural analysis process so as to connect the external loads to internal actions or stresses. Thus complex analyses such as finite element modelling may be involved.

5.5 Time-independent Structural Reliability

The solution of (11) may be sought through a number of options. The most obvious is to attempt to use numerical integration. This is not really feasible if the dimension of $\mathbf{X}$ is greater than about 5. The next best option is to use Monte Carlo simulation to perform the integration. However, in its most obvious and elementary form this is a highly inefficient procedure. Fortunately, considerable improvements can be made. One of these will be described below.

The alternative approach is to simplify the form of (11). This can be done through (i) simplifying the failure domain D_f to be a linear function (i.e. a 'first order' approximation) and (ii) simplifying $f_{\mathbf{X}}(\,)$, usually to a multi-normal (or -lognormal) form. These two simplifications allow the problem of integration to be by-passed altogether since simple rules can be used for the addition of random variables represented by their first and second moments (mean and variance). For obvious reasons, this approach is called the First Order Second Moment (FOSM) method. Because of its simplicity, it is one of the most well-known and widely used methods. It will be described below. However, unlike Monte Carlo methods, which produce accurate results in the limit, the FOSM method remains approximate. Various attempts have been made to improve it. These are also described, briefly.

5.6 Monte Carlo simulation

5.6.1 Basic Concepts

If we introduce an 'indicator' function $I[\,]$ defined such that $I[\theta]=1$ if the expression represented by θ is 'true' and zero otherwise, expression (11) can be rewritten as

$$p_f = J = \int \ldots \int I[G(\mathbf{x}) < 0] f_{\mathbf{X}}(\mathbf{x})\, d\mathbf{x} \qquad (12)$$

Evidently, there is a contribution to the integral only if the random vector $\mathbf{X}$ takes on values such that $I[\,]$ is unity, which means that the limit state(s) must be violated. Hence the indicator function in the integrand plays the same role as the limits of integration in (11). In (12) the limits of integration are over the whole range of $\mathbf{X}$ (or some reasonable approximation to it).

Expression (12) immediately provides the key to the application of simulation techniques. It can be written in a discrete form as

$$p_f \approx J_1 = \frac{1}{N} \sum_{j=1}^{N} I[G(\hat{\mathbf{x}}_j) \le 0] \qquad (13)$$

where $\hat{\mathbf{x}}_j$ represents a (the j th) vector of random samples selected from $f_{\mathbf{X}}(\,)$. These are known also as 'random deviates'. They can be obtained from appropriate sub-routines on many computers or can be generated from a set of random numbers. Details need not be considered here.

The application of (13) is simple. A discrete set of values $\hat{\mathbf{x}}_j$ is chosen for the random variables involved in the problem. Normally they would be generated using a computer. For each such set, the limit state function(s) is evaluated. Only if the structure fails, is the sum in (13) incremented by one. This process is repeated N times. The number of failures is then divided by the total number of times the structure was checked for safety (N). The result is the estimated probability of failure.

Sampling theory shows that (13) is an unbiased estimator of (12) and that as the number of samples taken increases, i.e. as $N \rightarrow \infty$, the approximation in (13) improves. In fact, the variance associated with the estimate may be estimated also and this is often useful in giving an indication of the accuracy of the solution.

Unfortunately, this process is extremely inefficient for high reliability structures as very few failures will be obtained. Several improvements have been suggested. The most well-known is 'importance sampling' (see below) but Latin hyper-cube sampling, antithetic variables, stratified sampling and others have been used as well. The reader should consult specialist texts for further information about these other techniques.

5.6.2 Importance sampling

In importance sampling (12) is rewritten as

$$p_f = J = \int \ldots \int I[G(\mathbf{x}) < 0] \frac{f_{\mathbf{X}}(\mathbf{x})}{h_{\mathbf{V}}(\mathbf{x})} h_{\mathbf{V}}(\mathbf{x}) d\mathbf{x} \qquad (14)$$

where $h_{\mathbf{V}}()$ is an 'importance sampling' probability density function. It can be selected (almost arbitrarily) by the investigator - more comments are given below. First we note that as a result, (13) becomes:

$$p_f \approx J_2 = \frac{1}{N} \sum_{j=1}^{N} \left\{ I[G(\hat{\mathbf{x}}_j) \leq 0] \frac{f_{\mathbf{X}}(\hat{\mathbf{v}}_j)}{h_{\mathbf{V}}(\hat{\mathbf{v}}_j)} \right\} \qquad (15)$$

which shows that the indicator function is now weighted to allow for the fact that the sample is taken from $h_{\mathbf{V}}()$ rather than from $f_{\mathbf{X}}()$. It should be evident that good choices of $h_{\mathbf{V}}()$ can considerably reduce the number of sample vectors $\hat{\mathbf{v}}_j$ required to produce a good estimate of p_f.

There has been considerable discussion about sensible choices for $h_{\mathbf{V}}()$ - its location and its shape. It is clear, however, that for general purposes a robust choice for $h_{\mathbf{V}}()$ places its mean at the point of maximum likelihood in the failure domain. This usually corresponds closely to the 'checking' or 'design' point in FOSM theory - see below. But these may not be known to the analysts and a useful guide is to select the point in $\mathbf{x}$ space (i.e. that combination of variables) for which the structure is considered most likely to fail. It has been shown that there is a reasonable degree of latitude in estimating this point without having very much effect on the estimate of p_f. For complex structures candidate checking points are more difficult to select intuitively and some form of formalized search or optimization routine may need to be employed. Various suggestions for this have been made.

For simplicity it is often sufficiently accurate to use a multi-normal distribution with independent components for $h_V()$, with standard deviations selected at 1-2 times those for the corresponding values for the components of $\mathbf{X}$. Where there are multiple candidate checking points $h_V()$ should be multi-modal with a mode corresponding to each checking point. Again, various suggestions have been made.

5.7 First Order Second Moment (FOSM) Theory

As introduced above, the approach used with the FOSM method to solve (11) is to simplify it. This can be done by (i) simplifying the failure domain D_f to be a linear function (i.e. a first order approximation) and (ii) simplifying $f_X()$ to a multi-normal (or -lognormal) form so that all the random variables are represented only by their first and second moments.

5.7.1 Simplest Case of FOSM

Consider again the simple bar under a tension load Q, as shown in Figure 5. When the load is applied the strength R of the bar may be sufficient to support the load or it may not be. Clearly, if the load is greater than the resistance of the bar it will fail. As before, we can define a 'safety margin' as

$$Z = R - Q \tag{16}$$

Here the failure condition is expressed by $Z < 0$ and the survival condition by $Z \geq 0$. Note that this is a linear limit state function. If the load and resistance are defined in terms of their second moment representations only, that is by their means μ and variances σ^2, we have, from probability theory rules (since R and Q are independent):

$$\mu_Z = \mu_R - \mu_Q \tag{17a}$$

$$\sigma_Z^2 = \sigma_R^2 + \sigma_Q^2 \tag{17b}$$

Hence the probability of failure becomes:

$$p_f = \text{Prob}(R - Q < 0) = \text{Prob}(Z < 0) = \Phi\left(\frac{0 - \mu_Z}{\sigma_Z}\right) = \Phi(-\beta) \tag{18}$$

where $\Phi(\,)$ is the well-known standard normal distribution function (with zero mean and unit standard deviation or variance). It is extensively tabulated in statistics texts, at least for higher probability levels. For the low values of probability usually associated with structural failure more detailed tables are required (e.g. Melchers, 1999).

Expression (18) can be visualized as shown in Figure 7 with the failure probability shown as the negative region. The parameter β in (18) is known as the 'safety index' or the 'reliability index'. Evidently, $\beta = \mu_Z / \sigma_Z$ and measures, in the space of the safety margin (see Figure 7) the distance from the mean of the safety margin to the failure condition in terms of the uncertainty σ_Z of the safety margin.

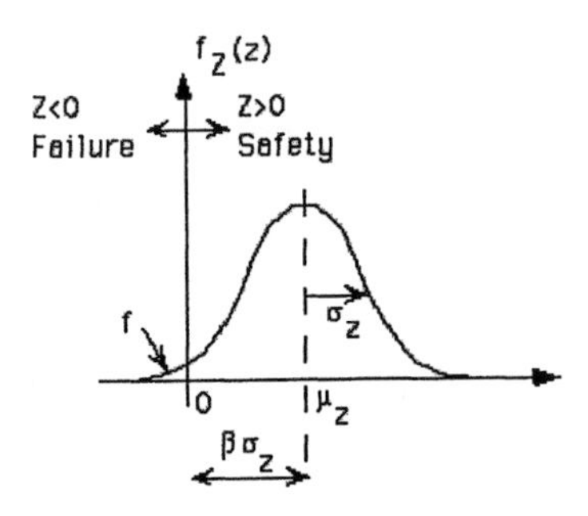

Figure 7 - Probability of failure and safety index.

It follows that if the uncertainty in either the applied load or in the resistance of the bar decreases, the value of σ_Z reduces and for given mean values of load and resistance (and hence of the safety margin) the safety index will be greater. Reference to Figure 7 will show that this means that the probability of failure will be less, as might be expected. As a result, greater β implies a lower probability of failure and vice-versa.

5.7.2 Multi-dimensional Case

The above concepts can be carried over directly to problems involving multiple resistance parameters and multiple loads (but not load processes - for these time-dependent theory must be used). Again, all random variables are described only by their mean and variance. It is conventional to transform them all to the standard normal space **Y** (with zero mean, unit variance). This means that the limit state function, which must be linear, will be transformed also. The transformation for each independent random variable X_i will be of the form:

$$Y_i = \frac{X_i - \mu_{X_i}}{\sigma_{X_i}} \tag{19}$$

with obvious notation. This transformation will need to be modified to allow for correlation where there is dependence between the variables or where they are not normal. In the latter case the Nataf, the Rosenblatt or some other transformation is required.

After transformation of the random variables the limit state function $G(\mathbf{x})=0$ becomes the function $g(\mathbf{y})=0$. It is not necessarily linear (as it must be for FOSM theory) and a first-order Taylor series expansion typically is applied to obtain the linear approximation. Without going in to detail, it can be shown that the most appropriate expansion point to be used for this is the checking point (and not the mean value point as used in some algorithms). Of course, at this stage the checking point is not yet known so that an iterative procedure will have to be employed for cases with non-linear $g(\mathbf{y})=0$.

For simplicity, assume now that $g(\mathbf{y})=0$ is linear. Then the problem can be sketched as in Figure 8 in two dimensions. In $\mathbf{y}$ space it shows the contours of the 'hill' described by the joint probability density function $f_{\mathbf{Y}}(\mathbf{y})$ of all the (transformed) random variables $\mathbf{Y}$. The probability of failure p_f is represented by the 'volume' under this hill in the failure region, that is the region for which $g(\mathbf{y})<0$. As before, rather than address p_f directly, it will be convenient to work with the safety index β.

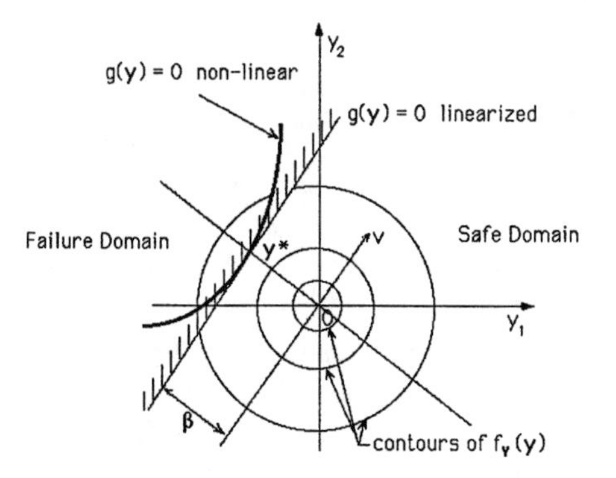

Figure 8 - Space of standard normal variables and linearized limit state function.

Since $g(\mathbf{y})=0$ is linear, it is evident that (i) integration in the direction v parallel to $g(\mathbf{y})=0$ will produce exactly the probability density function shown in Figure 7, (ii) that $\beta\sigma_Z$ in Figure 7 has now become just β since all the standard deviations were made unity by transformation, and (iii) β represents the shortest distance from the checking point to the origin of $\mathbf{y}$ space. These observations link the present discussion with the intuitive previous case and also provide the central statement of the FOSM problem:

$$\beta=\min\left(\mathbf{y}^{\mathrm{T}}.\mathbf{y}\right)^{1/2}=\min\left(\sum_{i=1}^{n}y_i^2\right)^{1/2} \tag{20}$$

where the y_i represent the coordinates of any point on the limit state surface $g(\mathbf{y})=0$. The particular point for which (20) is satisfied is the 'checking point' $\mathbf{y}^*$.

Clearly it is the projection of the origin on the (planar) limit state surface. This observation allows a number of geometric statements to be made and a popular iterative technique for finding $\mathbf{y}^*$ to be developed. These details will not be considered here.

5.7.3 Multiple limit states

From a practical point of view a more realistic situation is where there are several limit state functions which might apply to the structure. These do not cause difficulty with Monte Carlo methods but require some care with FOSM and its developments. Our discussion here will be conceptual rather than detailed.

A structure with several limit states can fail if any one of them is violated (a 'series' system) or if more than one must be violated for the structure to fail (a 'parallel' system for the limit states directly involved). A chain is a typical example of the first. A plastic collapse mode for a rigid frame, with each plastic hinge having a limit state function, is an example of the second. Of course, occurrence of any one plastic collapse mode will indicate failure of the structure, so that the collection of plastic failure modes represents a series system. Because of their importance, only series systems will be considered here.

A series system with m limit state functions can be represented in standard normal space $\mathbf{y}$ as shown in Figure 9. The failure region D_f is now that defined by the union of all the failure regions, shown shaded. Evidently, there will be a safety index β_i associated with each limit state function. Let the corresponding failure probability be $p_{fi} = \Phi(-\beta_i)$ for the ith limit state. This can be calculated directly for a single limit state using the procedure outlined above. Since the calculation of the 'volume' defined by the shaded region is not trivial, an approximation to it can be achieved using 'series bounds'. There are several of these, the simplest (and least accurate) being:

$$\max_{i=1}^{m}\left(p_{fi}\right) \le p_{f\,system} \le 1-\prod_{i=1}^{m}\left(1-p_{fi}\right) \approx \sum_{i=1}^{m} p_{fi} \tag{21}$$

with $p_{f\,system}$ denoting the probability of failure for the structural system with m limit state functions.

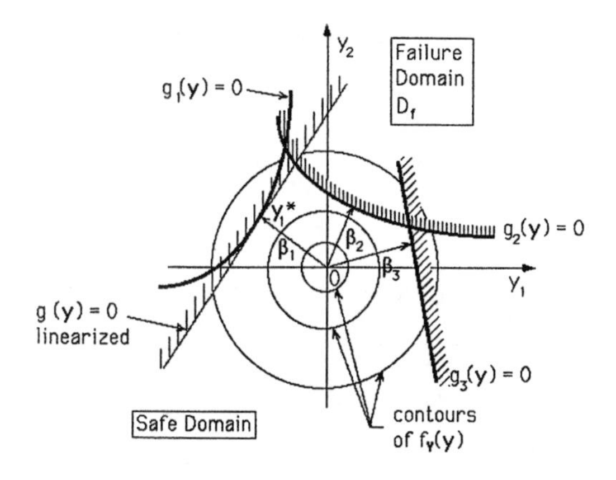

Figure 9 - Series system representation in standard normal space.

Developments of the above idea include more complex expressions for the bounds, considering various intersections of two (or more) limit state functions to refine the estimate of $p_{f\,system}$, and methods to eliminate one or more of the limit state functions if they can be shown to have insignificant influence on $p_{f\,system}$. These last methods are known as 'un-zipping' or 'truncated enumeration' or 'branch and bound' methods in the literature. They must be used with care, however, as there is no clear guarantee that important failure modes have not been eliminated. They also do not apply for situations with more than one load system, for the same reasons noted earlier.

5.8 First Order Reliability method (FORM)

In this section we will stay with linearized limit state functions (hence the 'first order' label) but now attempt to remove the limitation of second moment approximations being used for all the random variables. The approach used is quite simple, although the mathematical manipulation becomes involved, particularly for dependent variables. Fortunately, computer programs are now available. It will be sufficient, therefore, to review the main concept only.

The key observation to refining the FOSM method is that the probability of interest lies a long way from the origin in **y** space. Thus only the 'tails' of each of the probability distributions are of interest. The regions around the mean, while usually of great interest to statisticians, are of almost no interest in structural reliability theory. The approach adopted in the FORM method is to transform the relevant tail of a non-normal probability density function by the tail of normal one. This is shown schematically in Figure 10. Once the transformation is made, the substitute normal distribution is used in all computations, using FOSM techniques.

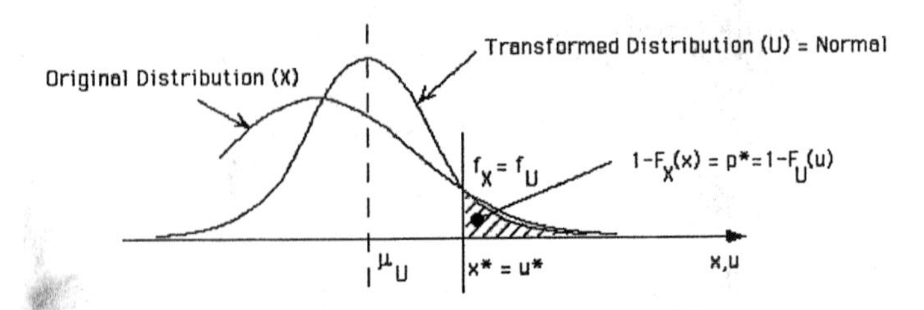

Figure 10 - Original and transformed normal distributions for the tail region.

The critical question is how the substitute normal distribution should be determined. Several approaches are possible. The simplest is to prescribe that for a given checking point $\mathbf{x}^*$ in the original problem space (i) the local probability densities at $\mathbf{x}^*$ shall be the same - i.e. $f_X(\mathbf{x}^*) = f_U(\mathbf{x}^*)$ and (ii) the probability p^* in the tails shall be the same - i.e. $1 - F_X(\mathbf{x}^*) = p^* = 1 - F_U(\mathbf{x}^*)$, see Figure 10. Of course, the checking point $\mathbf{y}^*$ in $\mathbf{y}$ space and hence $\mathbf{x}^*$ in $\mathbf{x}$ space is not known at the time the normal tail transformation must be made, so that it is clear that a 'trial-and-error' and iterative approach will need to be employed. Despite this disadvantage, the procedure works well.

5.9 Non-linear Limit States - Second Order Methods and Response Surfaces

The central pillar of the FOSM method and its extension to the FOR technique is the linear limit state function. Of course, in multi-dimensional space this is a hyper-plane but the equivalents to being able to integrate parallel to the plane (cf. direction v in Figure 8) and to being able to estimate the perpendicular distance β for the plane using (20), remain.
With non-linear limit state functions expression (20) can still be used, but the meaning of β as a measure of probability must then be modified to allow for the shape of the limit state surface. The second order approach to this is to let (in the case of a two-dimensional problem as in Figure 8) the actual limit state function be approximated by a second-order function (a quadratic or parabolic function). It is the possible to develop a relationship between the curvature of this function at the checking point and the correction to be made for the probability content between the second-order limit and the linear limit state functions.

For more dimensions the same general concept applies. The second order functions now become second-order surfaces. Such a surface is a special case of a 'response surface'. An extensive literature exists for the construction of such surfaces. Curvatures (or some equivalent measures) now need to be estimated in all principal directions and this may represent a significant computational issue. One alternative approach is to use Monte Carlo sampling to estimate the required correction between the non-linear surface and the approximating tangent hyper-plane of FOSM theory. Since interest lies in estimating a hyper-volume around the checking point, this type of sampling is much more efficient than that required to estimate the failure probability directly.

Response surfaces may be obtained by interpolation from discrete evaluations of complex and implicit limit states such as might be obtained as output from finite element or other numerical system analyses. Such problems still represent considerable challenges for structural reliability theorists. Although in principle it is possible to tie the FOSM method and refinements thereof to complex numerical analyses, there remains an unresolved balance between accuracy in reliability evaluation and in computational demands.

5.10 Modelling of Loads and Resistance

A structural reliability analysis depends critically on the information on which it is based. This applies to the limit state functions (and hence on structural analysis) but also on the modelling of the applied loads and the resistance parameters, including aspects such as dimensions and time-dependent properties. Curiously, perhaps, these matters have long been given rather scant attention by structural engineers, often having been content to rely on others to specify appropriate wind loading, earthquake loading etc. and to leave to materials specialists the definition of material properties. Of course, construction adequacy also is very important in obtaining adequate structures. However, it would appear that construction supervision is receiving less attention from structural engineers than in the past, in part because of current legal liability implications. Structural reliability theory, however, makes very clear the link between structural system safety and performance and the various factors which impact on these outcomes.

Chapter 4 in this volume has already discussed the important matter of loadings on multi-story structures and the modelling of that loading for use in probability analysis. Matters of durability are described in Chapter 7 and Chapters 10-12 deal with construction aspects Although there is no detailed discussion in this volume of resistance properties, they are readily available in the literature.

5.11 Conclusion

This chapter has reviewed the basic principles of structural reliability analysis using probabilistic methods. The main techniques open for reliability computations were described.

It is noted that unlike Monte Carlo methods, which should produce accurate results in the limit, although sometimes at extremely high computational cost, the FOSM method is approximate by definition. The various developments to improve its accuracy involve additional computation and sometimes undefinable levels of approximation.
The methods discussed generally do not deal directly with problems caused principally by major human error, although in principle the same basic framework can be used for such analyses provided the required data is available. The same limitation applies to any other technique which attempts to quantify reliability.

For applications such as demanded by structural design code calibration these issues have not been of major consequence since comparative measures rather than absolutes have been sufficient. This may not be the case in the future.

Bibliography

Ang, A.H.-S. and Tang, W.H. (1975) Probability Concepts in Engineering Planning and Design, Vol. II, Decision, Risk and Reliability, John Wiley & Sons.

Ditlevsen, O. and Madsen, H.O. (1996) Structural Reliability Methods, John Wiley & Sons.

Madsen, H.O., Krenk, S. and Lind, N.C. (1986) Methods of Structural Safety, Prentice-Hall.

Melchers, R.E. (1999) Structural Reliability Analysis and Prediction, Second Edition, John Wiley & Sons.

Schneider, J. (1997) Introduction to Safety and Reliability of Structures, (Structural Engineering Documents, No. 5) International Association for Bridge and Structural Engineering, Zurich.

Thoft-Christensen, P. and Baker, M.J. (1982) Structural Reliability and its Applications, Springer-Verlag.

Chapter 6

RELIABILITY-BASED DESIGN

By:
Marios K. Chryssanthopoulos, University of Surrey, UK
Dan Frangopol, University of Colorado, USA

6.1 INTRODUCTION

The primary objective of planning, design and quality assurance is to ensure satisfactory performance of engineering structures during their intended service life. Design is an iterative process aimed at finding the best possible solution. A key aspect that allows progressive improvement is the ability to describe performance through a number of pertinent criteria. Experience plays an important role in defining and refining these criteria, although for innovative structures it is as important to rely on past experience as it is to adapt and modify it.

Performance criteria can be thought of at different levels, from general to specific. For example, at a general level we want structures to be safe, functional and sustainable. For design and other purposes, 'safety' may be quantified through a number of more specific requirements, e.g.

- Structures shall withstand extreme and normal actions occurring during their construction and anticipated use,

- Structures shall not be damaged by rare accidental actions to an extent disproportionate to the original cause.

These, in turn, may be refined further using a set of specified limit states which separate desired from undesired states of the structure [1]. Thus, the first requirement above, often referred to as an ultimate limit state requirement, includes:

- Loss of equilibrium
- Attainment of maximum (or ultimate) capacity
- Transformation into a mechanism
- Instability, etc.

Occurring in a part (critical section, member) of the structure or in the whole structure.

The second requirement, termed the structural integrity (or robustness) requirement, may be related to energy absorption capability, loss of ductility, as well as post-ultimate equilibrium states. Similarly, functionality may be related to serviceability limit state requirements (such as local damage, unacceptable deformations, excessive vibrations), whereas sustainability is a more recent concept for which detailed requirements are more likely to be worked out on a case by case basis. However, in general, "sustainability refers to the ability to maintain structures at some desired level of performance or to change their performance at some desired rate and direction" [2] with due regard to environmental stressors and other concerns.

The design and construction of structural systems is undertaken under conditions of uncertainty. In this respect, compliance with the performance criteria listed above, either at a general or at a specific level is realistically possible only in probabilistic terms. Absolute safety or functionality cannot normally be achieved using finite resources, so the real question in a society with competing priorities becomes "What is an acceptable probability of achieving a specified performance level during the structure's life?" If safety is what we have in mind, the above question can be posed, more directly, as "how safe is safe enough?"

Answers to the above questions must be sought in the right context. To give just two examples:

- Are we referring to a specific and/or unique structure or to a large population constructed according to limits specified in a code of practice?

- Within a time framework, is the structure at the design stage, under construction or in use?

The purpose of this chapter is to summarize the principles and procedures used in reliability-based design. Starting from limit state concepts and their application to codified design, the link is made between unacceptable performance and probability of failure. This is then developed further in terms of a code format, in order to identify the key parameters and how they can be specified through probabilistic methods and reliability analysis. Emphasis is placed on new construction rather than existing structures, although some of the issues are common. In the latter parts of the chapter, recent concepts and applications pertaining to optimisation and minimum life-cycle cost are briefly presented.

6.2 DETERMINISTIC AND PROBABILISTIC DESIGN CONCEPTS

6.2.1 Limit States

As mentioned above, the performance of a whole structure or part of it may be described with reference to a set of limit states which separate acceptable states of the structure from unacceptable states. The boundary between acceptable (safe) and unacceptable (failure) states may be distinct or diffuse but, at present, codes of practice assume the former.

Thus, verification of a structure with respect to a particular limit state may be carried out via a model describing the limit state in terms of a function (called the limit state function) whose value depends on all design parameters. In general terms, attainment of the limit state can be expressed as

$$g(\mathbf{X}) = 0 \tag{6.1}$$

where **X** represents the vector of design parameters (also called the basic variable vector) that are relevant to the problem, and $g(\mathbf{X})$ is the performance function. Conventionally, $g(\mathbf{X}) \leq 0$ represents failure, i.e. an adverse state.

Basic variables comprise actions and their effects, material properties, geometrical data and factors related to the models used for constructing the limit state function. In many cases, important variations exist over time (and sometimes space), which have to be taken into account in specifying basic variables. In probabilistic terms, this may lead to random process rather than random variable models for some of the basic variables. However, simplifications might be acceptable, thus allowing the use of random variables whose parameters are derived for a specified reference period (or spatial domain) - see Chapter 5.

For many structural engineering problems, the performance function, $g(\mathbf{X})$, can be separated into one resistance function, $g_R(.)$, and one loading (or action effect) function, $g_S(.)$, in which case equation (6.1) can be expressed as

$$g_R(\mathbf{r}) - g_S(\mathbf{s}) = 0 \tag{6.2}$$

where **s** and **r** represent sub-sets of the basic variable vector, usually called loading and resistance variables respectively.

6.2.2 Partial Factors and Code Formats

Within present limit state codes, loading and resistance variables are treated as deterministic. The particular values substituted into equations (6.1) or (6.2) -the design values- are based on past experience and, in some cases, on probabilistic modelling and reliability calibration.

In general terms, the design value, x_{di}, of any particular variable is given by

$$x_{di} = \gamma_i \, x_{ki} \tag{6.3a}$$

$$x_{di} = x_{ki} / \gamma_i \tag{6.3b}$$

where x_{ki} is a characteristic (or representative) value and γ_i is a partial factor. Eqn (6.3a) is appropriate for loading variables whereas eqn (6.3b) applies to resistance variables, hence in both cases γ_i has a value greater than unity. For variables representing geometric quantities, the design value is normally defined through a sum (rather than a ratio), i.e. $x_{di} = x_{ki} \pm \Delta x$, where Δx represents a small quantity.

A characteristic value is strictly defined as the value of a random variable which has a prescribed probability of not being exceeded (on the unfavourable side) during a reference period. The specification of a reference period must take into account the design working life and the duration of the design situation.

The former (design working life) is the assumed period for which the structure is to be used for its intended purpose with maintenance but without major repair. Although in many cases it is difficult to predict with sufficient accuracy the life of a structure, the concept of a design working life is useful for the specification of design actions (wind, earthquake, etc.), the modelling of time-dependent material properties (fatigue, creep) and the rational comparison of whole-life costs associated with different design options. In Euro-code 1 [3], indicative design working lives range between 10 to 100 years, the two limiting values associated with temporary and monumental structures respectively.

The latter design situation represents the time interval for that the design will demonstrate that relevant limit states are not exceeded. The classification of design situations mirrors, to a large extent, the classification of actions according to their time variation. Thus, design situations may be classified as persistent, transient or accidental. The first two are considered to act with certainty over the design working life. On the other hand, accidental situations occur with relatively low probability over the design working life. Clearly, whether certain categories of actions (snow, flood, and earthquake) are deemed to give rise to transient or accidental situations will depend on local conditions. Typically, the load combination rules are not the same for transient and accidental situations, and also a degree of local damage at ultimate limit state is more widely accepted for accidental situations. Hence, the appropriate load classification is a very important issue in structural design.

In treating time-varying loads, values other than the characteristic may be introduced. These so-called representative values are particularly useful when more than a single time-varying load acts on the structure. For material properties a specified or nominal value is often used as a characteristic value, and since most material properties are assumed to be time-independent, the above comments are not relevant. For geometrical data, the characteristic values usually correspond to the dimensions specified in design.

Partial factors account for the possibility of unfavourable deviations from the characteristic value, inaccuracies and simplifications in the assessment of the resistance or the load effect, uncertainties introduced due to the measurement of actual properties by limited testing, etc. The partial factors are an important element in controlling the safety of a structure designed according to the code but there are other considerations to help achieve this objective. Note that a particular design value x_{di} may be obtained by different combinations of x_{ki} and γ_i.

The process of selecting the set of partial factors to be used in a particular code can be seen as a process of optimization such that the outcome of all designs undertaken to the code is in some sense optimal. Such a formal optimization process is not usually carried out in practice; even in cases where it has been undertaken, the values of the partial factors finally adopted have been adjusted to account for simplicity and ease of use. More often, partial factor values are based on a long experience of building tradition. However, it is nowadays generally accepted that a code should not be developed in a way that contradicts the principles of probabilistic design and its associated rules.

Eqn (6.2), lends itself to the following deterministic safety checking code format

$$\gamma_{Sd} s(F_d, a_d, \ldots) \leq \frac{1}{\gamma_{Rd}} r(f_d, a_d, \ldots) \tag{6.4}$$

where F_d, f_d and a_d are design values of basic variables representing loading, resistance and geometrical variables respectively, which can be obtained from characteristic/representative values and associated partial factors, and γ_{Sd}, γ_{Rd} are partial factors related to modelling uncertainties (loading and resistance functions respectively). More general [1], as well as alternative code formats which nevertheless recognise and treat the same uncertainties may be found in the literature [3-8].

As can be seen, the safety checking equation controls the way in which the various clauses of the code lead to the desirable level of safety of structures designed to the code. It relates to the number of design checks required, the rules for load combinations, the number of partial factors and their position in design equations, as well as whether they are single or multiple valued, and the definition of characteristic or representative values for all design variables.

In principle, there is a partial factor associated with each variable. Furthermore, the number of load combinations can become large for structures subjected to a number of permanent and variable loads. In practice, it is desirable to reduce the number of partial factors and load combinations while, at the same time, ensuring an acceptable range of safety level and an acceptable economy of construction. Hence, it is often useful to make the distinction between primary basic variables and other basic variables. The former group includes those variables whose values are of primary importance for design and assessment of structures. The above concepts of characteristic and design values, and associated partial factors, are principally relevant to this group. Even within this group, some partial factors might be combined in order to reduce the number of factors. Clearly, these simplifications should be appropriate for the particular type of structure and limit state considered.

6.2.3 Structural Reliability

Load, material and geometric parameters are subject to uncertainties. They can be represented by random variables (this being the simplest possible probabilistic representation; as noted above, more advanced models might be appropriate in certain situations, such as random fields).

Thus, as presented in Chapter 5, the probability of occurrence of the failure event, P_f, is given by

$$P_f = \text{Prob}\{ g(\mathbf{X}) \leq 0 \} = \text{Prob}\{ M \leq 0 \} \quad (6.5a)$$

where, M = g (**X**) and **X** now represents a vector of basic random variables. Note that M is also a random variable, usually called the safety margin.

If the limit state function can be expressed in the form of equation (6.2), equation (6.5a) may be written as

$$P_f = \text{Prob}\{ r(\mathbf{R}) \leq s(\mathbf{S}) \} = \text{Prob}\{ R \leq S \} \quad (6.5b)$$

where R = r (**R**) and S = s (**S**) are random variables associated with resistance and loading respectively.

Using the joint probability density function of **X**, $f_{\mathbf{X}}(\mathbf{x})$, the failure probability defined in equation (6.5) can now be determined from

$$P_f = \int_{g(\mathbf{X}) \leq 0} f_{\mathbf{X}}(\mathbf{x}) d\mathbf{x} \quad (6.6)$$

Schematically, $g(\mathbf{X}) = 0$ represents the boundary between safety and failure, and the integration domain of equation (6.6) describes failure - see Chapter 5.

The reliability, P_s, associated with the particular limit state considered is

$$P_s = 1 - P_f \quad (6.7)$$

In recent years, a standard reliability measure, the reliability index β, has been adopted which has the following relationship with the failure probability

$$\beta = -\Phi^{-1}(P_f) = \Phi^{-1}(P_s) \quad (6.8)$$

where $\Phi^{-1}(.)$ is the inverse of the standard normal distribution function, see Table 6.1.

Table 6.1: Relationship between β and P_f

P_f	10^{-1}	10^{-2}	10^{-3}	10^{-4}	10^{-5}	10^{-6}	10^{-7}
β	1.3	2.3	3.1	3.7	4.3	4.7	5.2

The basis for this relationship has been outlined in Chapter 5. Here, it suffices to state that the reliability index β is often used in probability-based design as a surrogate quantity, i.e. *in lieu* of the probability of failure, although equation (6.8) is strictly valid under restrictive conditions. This is partly attributed to the evolution of structural reliability theory, and partly to the fact that it is often easier to quote β values (say in the range from 1 to 8) rather than P_f values (in the roughly equivalent range from 10^{-1} to 10^{-16}).

In practically all structural engineering applications, complete statistical information about the basic random variables **X** is not available, and the function $g(.)$ is a mathematical model which idealizes the limit state rather than represent reality. In this respect, the probability of failure is a point estimate given a particular set of assumptions regarding probabilistic modelling and a particular mathematical model for $g(.)$.

The uncertainties associated with these models can be represented in terms of a vector of random parameters Θ, and hence the performance function may be re-written as $g(\mathbf{X}, \Theta)$. It is important to note that the nature of uncertainties represented by the basic random variables **X** and the parameters Θ is different. Whereas uncertainties in **X** cannot be influenced without changing the physical characteristics of the problem, uncertainties in Θ can be influenced by the use of alternative methods and collection of additional data.

In this context, eqn (6.6) may be recast as follows

$$P_f = \int_{g(\mathbf{X},\Theta)\leq 0} f_{\mathbf{X}|\Theta}(\mathbf{x}\,|\,\theta)d\mathbf{x} \tag{6.9}$$

where $P_f(\theta)$ is the conditional probability of failure for a given set of values of the parameters θ and $f_{\mathbf{X}|\Theta}(\mathbf{x}|\theta)$ is the conditional probability density function of **X** for given θ [9].

In order to account for the influence of parameter uncertainty on failure probability, one may evaluate the expected value of the conditional probability of failure, i.e.

$$\overline{P_f} = \mathrm{E}\,[P_f(\theta)] = \int_{\theta} P_f(\theta)\, f_{\Theta}\,(\theta)\, d\theta \tag{6.10a}$$

where $f_{\Theta}(\theta)$ is the probability density function of Θ. The corresponding reliability index is given by

$$\overline{\beta} = -\,\Phi^{-1}(\overline{P_f}) \tag{6.10b}$$

The main objective of reliability analysis is to estimate the failure probability (or, the reliability index). Hence, it replaces the deterministic safety checking format (e.g. equation (6.4)), with a probabilistic assessment of the safety of the structure (typically equation (6.6) but also in few cases equation (6.9)). Depending on the nature of the limit state considered, the uncertainty sources and their implications for probabilistic modelling, the characteristics of the calculation model and the degree of accuracy required, an appropriate methodology has to be developed. In many respects, this is similar to the considerations made in formulating a methodology for deterministic structural analysis but the problem is now set in a probabilistic framework.

A fundamental set of limit states is that for which all the variables are treated as time independent, either by neglecting time variations in cases where this is considered acceptable or by transforming time-dependent processes into time-invariant variables. For these problems, the First Order Second Moment theory (see Chapter 5) is often used for reliability computation. In this method, reliability is estimated by determining a suitable checking point in standard normal space; as shown in Chapter 5 this turns out to be the point on the limit-state surface which is closest to the origin, and is called the 'design point'. In standard normal space it represents the most likely failure point, in other words its co-ordinates define the combination of variables that are most likely to cause failure. Thus, following the notation of chapter 5, the reliability index is given by

$$\beta = \sqrt{\sum_{i=1}^{n} y_i^{*2}} \tag{6.11}$$

where the vector $\mathbf{y}^*$ denotes the co-ordinates of the design point, i.e.

$$\mathbf{y}^* = (y_1^*, y_2^*, \dots y_n^*)$$

Note that $\mathbf{y}^*$ can also be written as

$$\mathbf{y}^* = \beta \alpha \tag{6.12}$$

where $\alpha = (\alpha_1, \alpha_2, \dots \alpha_n)$ is the unit normal vector to the limit state surface at $\mathbf{y}^*$, and, hence, α_i (i=1,...n) represent the direction cosines at the 'design point'. These are also known as the sensitivity factors, as they provide an indication of the relative importance of the uncertainty in basic random variables on the computed reliability. Their absolute value ranges between zero and unity and the closer this is to the upper limit, the more significant the influence of the respective random variable is to the reliability.

6.2.4 Interpretation of Reliability Analysis Results

As mentioned above, under certain conditions the 'design point' in standard normal space, and its corresponding point in the basic variable space, is the most likely failure point. Since the objective of a deterministic code of practice is to ascertain attainment of a limit state, it is clear that any check should be performed at a critical combination of loading and resistance variables and in this respect, the 'design point' values from a reliability analysis are a good choice. Hence, in the deterministic safety checking format, equation (6.4), the design values can be directly linked to the results of a reliability analysis, i.e. P_f or β and α_i's. Thus, the partial factor associated with a basic random variable X_i, is determined as

$$\gamma_{Xi} = \frac{x_{di}}{x_{ki}} = \frac{F^{-1}_{Xi}(\Phi(u_i^*))}{x_{ki}} = \frac{F^{-1}_{Xi}(\Phi(\alpha_i\beta))}{x_{ki}} \quad (6.13)$$

where x_{di} is the 'design point' value and x_{ki} is a characteristic value of X_i, As can be seen, the 'design point' value can be written in terms of the original distribution function $F_{Xi}(.)$, the reliability analysis results, i.e. β and α_i, and the standard normal distribution function $\Phi(.)$.

If X_i is normally distributed, equation (6.13) can be written (after non-dimensionalising both x_{di} and x_{ki} with respect to the mean value) as

$$\gamma_{Xi} = \frac{1 - \alpha_i \beta v_{Xi}}{1 + k v_{Xi}} \quad (6.14)$$

where v_{Xi} is the coefficient of variation and k is a constant related to the fractile of the distribution selected to represent the characteristic value of the random variable X_i. As shown, equations (6.13) and (6.14) are used for determining partial factors of loading variables, whereas their inverse is used for determining partial factors of resistance variables. Similar expressions are available for variables described by other distributions (e.g. log-normal, Gumbel type I, given in, for example, [3]). Thus, partial factors could be derived or modified using results from First Order Second Moment reliability analysis. If the reliability assessment is carried out using simulation, sensitivity factors are not directly obtained, though, in principle, they could be through some additional calculations.

6.3 RELIABILITY-BASED CODE CALIBRATION

It is evident from equations (6.13) and (6.14) that the reliability index β can be linked directly to the values of partial factors adopted in a deterministic code. Thus, early applications of reliability theory exploited this link in order to undertake reliability-based code calibration. This process coincided with the move in the structural engineering community from 'working stress' to 'limit state' codes. It has been actively pursued since the 1970's in relation to US [4-6], European (e.g. [7, 8]) and Far Eastern codes and is described in a number of standard reliability textbooks, e.g. [9-12]. A critical review of reliability-based codified design has been presented by Ellingwood [13]. As pointed out therein, some key issues that remain following these calibration studies include:

- Target reliability levels were derived by designing a large number of relatively simple members, e.g. beams, columns, tension members and connections, to the limit of the earlier design code, thus enforcing, on average, calibration to existing practice. Essential though this was for continuity, it allowed inefficiencies and inconsistencies to propagate in the new codes. It also virtually completely overlooked the issue of component versus system failure, which can be very important in complex, possibly redundant, structures; this point is revisited in section 6.5 below.

- Calibrated reliabilities were treated as 'notional' values, with no attempt made to rationalize them in explicit risk terms; this can partly be attributed to how many novel features could the generally conservative structural engineering community accept in one cycle, but is also linked to the 'tail sensitivity' problem (see Chapter 5) and the real difficulty of backing up these 'notional' values with comparable accident statistics.

Notwithstanding the above limitations, the process of reliability-based code calibration has spearheaded numerous studies dealing with structural design issues in a probabilistic framework. Typically, such studies start by accepting a set of hypotheses, e.g. current deterministic code provisions, and produce a range of structural designs which conform to this set. Relevant limit states to describe possible failure modes (e.g. bending, shear, buckling etc) are formulated and probabilistic modelling of random variables is undertaken bearing in mind the context in which the study is performed (e.g. in-situ or pre-cast/pre-fabricated construction for geometry, dead load and material property modelling; structure location and exposure characteristics for live load modelling). Using the methods outlined in Chapter 5, reliability analysis of the various designs is then performed and results in terms of reliability index and sensitivity factors are obtained. From these results, a range of observations and comparative assessments is usually possible which can lead to the discovery of unfavourable features or undesirable trends invoked by accepting the original set of hypotheses. Improvements can thus be made on the basis of the results of the reliability analysis.

As mentioned above, this type of study, which effectively is a reliability-based design assessment, has become quite popular in the last twenty five years or so. Practically all basic structural members (tension, compression, bending) in the two most common construction materials (concrete, steel) have been the subject of such studies. In recent years, whole structural systems, e.g. steel or reinforced concrete frame buildings and bridges, have been examined in this manner.

It is clearly beyond the scope of the present chapter to produce a comprehensive review in this area. However, by way of example, the reader is directed to a few studies dealing with reinforced concrete members and frames from which the type of conclusions reached on the basis of probabilistic analysis might be glimpsed:

(i) In a study dealing with reinforced concrete members, Ellingwood [14] observed that code provisions at the time tended to produce beam designs which were more likely to fail in shear than in flexure; since the former is a failure mode with considerably less ductility than the latter, this was an example of an undesirable inconsistency in a deterministic code which was revealed through uncertainty modelling and reliability analysis.

(ii) Frangopol and his colleagues [15] studied the reliability of reinforced concrete columns, paying particular attention to load-path and load correlation effects; it is worth noting that correlation is often treated inappropriately in code provisions. For example, the study showed that the assumption of proportional loading used in most building codes for RC columns may be unconservative in the tension failure region.

(iii) Kappos et al [16] looked at the effect of uncertainty on the ductility of RC members, a property which is important for seismic applications. It was found that the use of code specified material property values leads to calculated ductilities that have a high probability of not being met; thus, suggestions for calibration of code provisions for ductility were made.

(iv) Design of buildings in seismic regions is critically dependant on the so-called design earthquake, which is based on probabilistic, and other, considerations. However, the reliability of the final designs remains, to a large extent, unquantified. Wen [17] has presented reliability evaluation and comparison of buildings designed in accordance with current code procedures in different countries, whereas Chryssanthopoulos et al [18] have estimated the probability distribution of the so-called 'behaviour' factor in RC frames designed to the limit of current Eurocodes.

6.4 RELIABILITY DIFFERENTIATION

At the same time as reliability calibration studies were being undertaken, and in many ways compelled by the need to interpret the results and convince the wider structural engineering community of their usefulness, a number of researchers tackled the fundamental issues in target reliability specification. Thus, it is now widely accepted that the appropriate degree of reliability should be judged with due regard to the possible consequences of failure and the relative costs of safety measures [1]; the latter is a function of the expense, level of effort and procedures necessary to reduce the risk of failure.

Using somewhat different terminology but essentially describing the same factors, 'the appropriate degree of reliability' should take into account the cause and mode of failure, the possible consequences of failure, the social and environmental conditions, and the cost associated with various risk mitigation procedures [1, 3-8]. For example, Euro-code 1 [3] contains an informative annex in which target reliability indices for ultimate limit states are given for three different reliability classes. Table 6.2 reproduces the recommended target reliability values from this document. The three reliability classes (RC1 to RC3) may be associated to three corresponding consequences classes (CC1 to CC3) which are defined as follows:

CC3: High consequence for loss of human life *or* economic, social or environmental consequences very great.

CC2: Medium consequence for loss of human life, economic, social or environmental consequences considerable.

CC1: Low consequence for loss of human life *and* economic, social or environmental consequences small or negligible.

ISO 2394 [1] contains a similar table, in which target reliability is linked explicitly to consequences of failure and the relative cost of safety measures. Other recently developed codes of practice have made explicit allowances for 'system' effects (i.e. failure of a redundant vs. non-redundant structural element) and inspection levels.

Table 6.2: Recommended target reliability indices in Structural Eurocodes [3]

Reliability Class	Target Value for β	
	1 year reference period	50 years reference period
RC3	≥ 5.2	≥ 4.3
RC2	≥ 4.7	≥ 3.8
RC1	≥ 4.2	≥ 3.3

Different target values would be adopted for serviceability limit states or for limit states associated with robustness requirements. The former should be compatible with the consequences associated with loss of functionality, whereas the latter should take into account, for example, the time period for which the structure will need to perform after a degree of local damage is sustained.

6.5 FULL PROBABILISTIC DESIGN

As mentioned earlier, if certain conditions are satisfied, the reliability index provides a direct estimate of the probability of failure. In general, first and second-order reliability methods (FORM/SORM) must be used to find the reliability index. Both FORM and SORM are analytical probability integration methods using first and second-order approximations of the failure surface, respectively. Monte Carlo simulation approaches, such as crude simulation, variance reduction techniques, adaptive sampling, can be used to approximate the failure probability defined in equation (6.6). These approaches are more general but, in most cases, less efficient than FORM/SORM.

As indicated in Table 6.2, the fundamental quantitative measure of acceptable performance is the target reliability index. This index is a surrogate for the target failure probability. In the United States, the LRFD specifications for different materials and structures are all based on 'notional' reliabilities [4, 5]. As indicated by Ellingwood [13], "these notional reliabilities, encapsulated in reliability indices, were obtained by calibrating the proposed codes to traditional practice rather than from quantitative risk analysis." As mentioned earlier, this calibration allows inconsistencies in current practice to propagate into the new specifications. For this reason, the profession is now in a critical stage where direct reliability-based design (i.e. design based on 'notional' reliability indices) is a possibility that has to be investigated with caution. This is because current estimates of reliabilities "are not dependable enough to be used directly by individual structural engineers in design without validation" [13].

With the above considerations in mind, moving toward a full probabilistic design is a responsibility for the reliability research community. This move began about three decades ago with the pioneering work of Cornell [19] and Ang and Cornell [20]. A concerted effort is currently under way under the auspices of the Joint Committee for Structural Safety (JCSS) to develop an operational probabilistic code, which will include full modelling information on the random quantities that are encountered in structural design. In order to facilitate dissemination and to enable periodic updating on the basis of new information, the JCSS probabilistic model code has been available as a web-based publication since 2001 and is continuously being expanded and updated [21].

In order to illustrate full probabilistic design, Figures 1 to 4 present results associated with normal and log-normal models of the basic variables R and S. Figures 1 and 2 show that under conditions of uncertainty, embedded in the coefficients of variation $V(R)$ and $V(S)$, it is possible to design for a given target reliability level β. The purpose of design is to determine the value of θ, which represents the ratio between the mean values of R and S, and, from that, the required mean value of R. Figure 3 shows that for a given target reliability level, the ratio of mean values can be vastly different depending on $V(R)$ and/or $V(S)$. Finally, Figure 4 indicates that under conditions of uncertainty a design based on a prescribed ratio of mean values results in vastly different reliability levels.

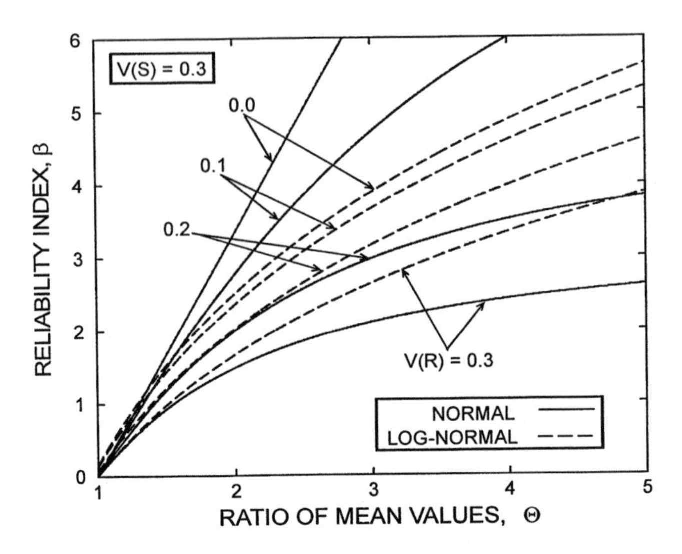

Figure 1. Variation in Reliability Index β due to Changes in Ratio of Mean Values θ and Coefficient of Variation V(R) for a Fixed Coefficient of Variation of Load Effect, V(S) =0.3

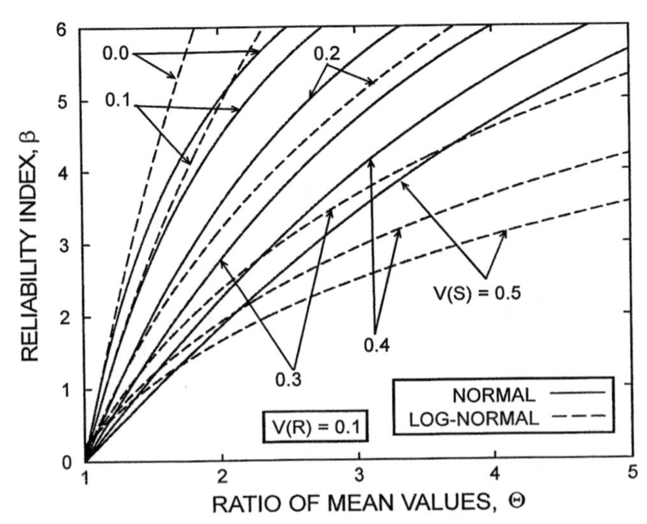

Figure 2. Variation in Reliability Index β due to Changes in Ratio of Mean Values θ and Coefficient of Variation V(S) for a Fixed Coefficient of Variation of Resistance, V(R) =0.1

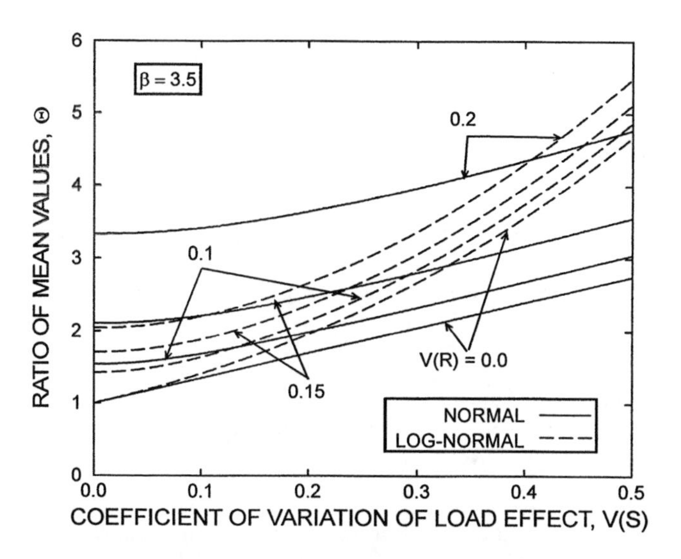

Figure 3. Variation of Ratio of Mean Values θ due to Changes in Coefficients of Variation V(S) and V(R) for a Fixed Reliability Index, $\beta = 3.5$

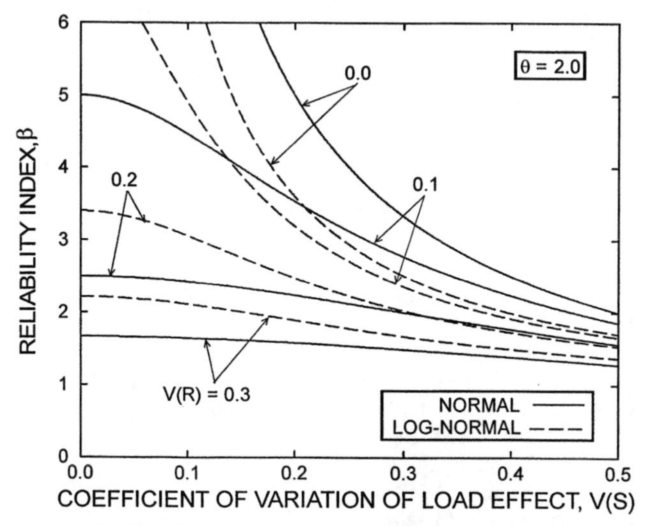

Figure 4. Variation of Reliability Index β due to Changes in Coefficients of Variation V(S) and V(R) for a Fixed Ratio of Mean Values, $\theta = 2.0$

In order to implement the full probabilistic approach in tall building design practice, further efforts should be made in developing target reliability levels for building components and systems. Also, issues such as system behavior, redundancy, ductility, aging and deterioration have to be carefully considered in this implementation [22-30].

6.6 LIFE-CYCLE RELIABILITY-BASED DESIGN

Life-cycle cost engineering treats all stages of life of a facility (i.e. design, construction, inspection, maintenance, rehabilitation and decommissioning) as part of an integrated process of evaluation by considering both cost and performance throughout the planning horizon of this facility. As indicated by Fisher [31], "consideration of all stages in the life-cycle brings to light future problems and issues that can occur downstream, and therefore, supports intelligent and informed decision-making so that overall life-cycle costs will be reduced".

In the design of civil infrastructure systems, including tall buildings, lifetime reliability and life-cycle cost are of main concern. Although this is generally recognized, the significance of life-cycle cost has not been properly integrated with the reliability issues in the development of design criteria for buildings [32].

At the ICOSSAR'97 panel discussion on life-cycle cost evaluation and target reliability level for design [33] and at the IABSE Congress 2000 [34] various aspects of the integration of life-cycle cost analysis with reliability analysis including optimum balance between safety and economy, and minimum expected cost of civil infrastructure systems were presented and discussed. In general, it was agreed that the incorporation of life-cycle cost requirements in the design process will improve the management of these systems so that the overall life-cycle cost will be reduced.

In life-cycle reliability-based design the total life-cycle cost can be formulated as [35]

$$C_{E,L} = C_I + C_{PM,L} + C_{INS,L} + C_{REP,L} + C_{FAIL,L} \tag{6.15}$$

where $C_{E,L}$ = total expected cost over the life span L of the structure (i.e. service life),
C_I = initial cost of the structure,
$C_{PM,L}$ = preventive maintenance cost over the life span of the structure,
$C_{INS,L}$ = inspection cost over the life span of the structure,
$C_{REP,L}$ = repair cost of the structure over its life span, and
$C_{FAIL,L}$ = $C_F\, P_{F,L}$ = expected failure cost of the structure
C_F = costs associated with failure of the structure
$P_{F,L}$ = probability of failure over the life span.

The optimization problem consists of minimizing the total expected cost under reliability constraints as follows:

$$min\ C_{E,L} \tag{6.16}$$

subject to

$$\beta_{L,i}(t) \geq \beta^{*}_{L,i} \tag{6.17}$$

where $\beta_{L,i}(t)$ and $\beta^{*}_{L,i}$ are the time-dependent reliability index and target reliability index for limit state i (associated with the life span L of the structure), respectively. For example, if the total cost is defined in the simplified format given by equation (4.13), see section 4.6, the target value $\beta^{*}_{L,i}$ can be obtained from Figure 1.

The total cost of a building over its life-cycle consists of initial, inspection, maintenance (including preventive maintenance), repair and failure costs. To perform life-cycle cost analysis, the time value of money must be considered. Thus, costs have to be discounted to present values, C_{PV}, using the expression [36]

$$C_{PV} = \frac{C}{(1+r)^t} \tag{6.18}$$

where C is cost at current price,
r is the discount rate,
t is the time period in years, and
the term $1/(1+r)^t$ is the discount factor.

The choice of the discount rate is mainly a political decision serving as an agreement on comparing investments. Under the same economical situation, a government agency can impose a different discount rate than that used by a private entity. The investment with minimum expected discounted cost over a bounded time horizon (say, 30 years) is often preferred. There are also models considering an unbounded time horizon for decision-making [37]. The advantage of these models is that, for relatively large time horizons (say, larger than 60 years), the expected discounted cost can be computed by using a simple analytical formula provided by the renewal theory. The current value of discount rate used in different countries varies from 2% to 10%. As pointed out above, the discount rate has significant implications for the management of structures.

The non-linear optimization problem defined in equations (6.16) and (6.17) can be solved as indicated by Frangopol *et al.* [38]. However, there are many aspects of management of structures based on balancing reliability and life-cycle cost that are currently under investigation [13, 27, 33, 38-50]. These aspects include the accuracy of predictive damage models and their effect on structural reliability, the trade-off between the extent and accuracy of inspections, the quantification of the effects of various maintenance interventions, i.e. essential, preventive, and mixed (essential and preventive), on structural reliability, the required reliability level, and the quantification of failure cost and its impact on the optimal solution. Efforts should be increased to provide the data necessary for the practical application of life-cycle reliability design in order to obtain the optimum balance between lifetime reliability and life-cycle cost.

6.7 CONCLUDING REMARKS

It is clear from the above that reliability-based design has made significant strides in the last thirty years or so. Elementary component reliability formulations have progressively been substituted with more advanced approaches, sometimes at a system level. Moreover, methodologies dealing with life-cycle cost optimisation are currently being actively researched and promoted by a number of relevant authorities. Finally, in a number of codes and background documents target reliability values have now been put forward, whereas reliability differentiation has moved from a pure concept to an applied tool.

That said, considerable challenges remain if reliability-based design is to be widely accepted within the structural engineering community. These include:

- The establishment of a link between 'notional' failure probabilities and field accident statistics. In this context, it is worth emphasizing that a large proportion of structural failures can be attributed to human error. The reliability framework outlined in this chapter does not, on its own, provide the means to guard against human error. An overview of the types of human error in construction, and how these can be minimised is the subject of Chapter 12.
- The continuing need to support the selection of basic variable distributions with real life data.
- The advances required in order to define or refine performance-based criteria for complex structural systems, such as tall buildings.
- The need to collect data and, thereby, improve the reliability of cost models in life-cycle cost approaches.

Clearly the list is not exhaustive but merely indicative of some important issues. Efforts to resolve them will undoubtedly help in achieving the next generation of reliability-based designs for structures.

REFERENCES

/1/ International Organization for Standardization (ISO), *General Principles on Reliability for Structures*, ISO 2394, 1998.

/2/ Zimmerman R and Sparrow R, *Workshop on Integrated Research for Civil Infrastructure Systems,* July 15-17, 1996, Washington D.C., *Final Report to the National Science Foundation,* New York University, New York, 1997.

/3/ European Standard, EN 1990: 2002, *Eurocode: Basis of Structural Design*, European Committee of Standardisation CEN/TC250, Brussels, 2002.

/4/ Galambos T V, Ellingwood B, MacGregor J G and Cornell C A, Development of Probability-based Load Criteria: Assessment of Current Design Practice, *J. Str. Eng., ASCE*, 108, 1982, pp. 959-977.

/5/ Ellingwood B, MacGregor J G, Galambos T V and Cornell C A, Development of Probability-based Load Criteria: Load Factors and Load Combinations, *J. Str. Eng., ASCE*, 108, 1982, pp. 978-997.

/6/ Ellingwood B and Galambos T V, Probability Based Criteria for Structural Design, *Structural Safety*, 1, pp. 15-26.

/7/ Comité Européen du Béton (CEB), First Order Concepts for Design Codes, CEB Bulletin No. 112, Munich, 1976.

/8/ Construction Industry Research and Information Association (CIRIA), *Rationalisation of Safety and Serviceability Factors in Structural Codes,* Report 63, London, 1977.

/9/ Der Kiureghian A, Measures of Structural Safety under Imperfect States of Knowledge, *J. Str. Eng., ASCE*, 115 (5), 1989, pp. 1119-40.

/10/ Thoft-Christensen P and Baker M J, *Structural Reliability Theory and its Applications*, Springer-Verlag, 1982.

/11/ Ditlevsen O and Madsen H O, *Structural Reliability Methods,* J Wiley, 1996.

/12/ Melchers R E, *Structural Reliability Analysis and Prediction,* 2nd edn, J Wiley, 1999.

/13/ Ellingwood B R, Acceptable risk bases for design of structures, *Progress in Structural Engrg. and Mat.*, 3(2), 2001, pp.170-179.

/14/ Ellingwood B R, Reliability of Current Reinforced Concrete Designs, *J. Str. Div., ASCE*, 105 (ST4), 1979, pp. 699-712.

/15/ Frangopol D M, Ide Y, Spacone E and Iwaki I, A New Look at Reliability of Reinforced Concrete Columns, *Structural Safety,* 18, 1996, pp. 123-50.

/16/ Kappos A J, Chryssanthopoulos M K and Dymiotis C, Uncertainty Analysis of Strength and Ductility of Confined Reinforced Concrete Members, *Engineering Structures*, 21, 1999, pp. 195-208.

/17/ Wen Y K, Building Reliability and Code Calibration, *Earthquake Spectra*, 11/2, 1995, pp. 269-296.

/18/ Chryssanthopoulos M K, Dymiotis C and Kappos A J, Probabilistic Calibration of Behaviour Factors in EC8-Designed R/C Frames, *Engineering Structures*, 22, 2000, pp. 1028-1041.

/19/ Cornell C A, A probability-based structural code, *Journal of the American Concrete Institute*, 66 (12), 1969, pp. 974-985.

/20/ Ang A H-S and Cornell C A, Reliability bases of structural safety and design,*Journal of Structural Engineering*, ASCE, 100 (9), 1974, pp. 1755-1769.

/21/ The JCSS Probabilistic Model Code, http://www.jcss.ethz.ch.

/22/ Frangopol D M and Curley J P, Effects of damage and redundancy on structural reliability, *Journal of Structural Engineering*, ASCE, 113 (7), 1987, pp. 1533-1549.

/23/ Fu G and Frangopol D M, Balancing weight, system reliability and redundancy in a multiobjective optimization framework, *Structural Safety*, 7(2-4), 1990, pp. 165-175.

/24/ Moses F, New directions and research needs in system reliability research, *Structural Safety*, Elsevier, 7(1), 1990, pp. 93-108.

/25/ Hendawi S and Frangopol D M, System reliability and redundancy in structural design and evaluation, *Structural Safety*, Elsevier, 16(1+2), 1994, pp. 47-71.

/26/ Bertero R D and Bertero V V, Redundancy in earthquake-resistant design, *Journal of Structural Engineering*, ASCE, 125 (1), 1999, pp. 81-88.

/27/ Ellingwood B R, Reliability–based structural design: Current status and challenges, in *Proceedings of Asian–Pacific Symposium on Structural Reliability and its Applications APSSRA99*, Taipei, Taiwan, 1999, pp. 63-74.

/28/ Wen Y K, Structural redundancy of steel buildings under seismic loads, in *Proceedings of Asian – Pacific Symposium on Structural Reliability and Its Applications*, Taipei, Taiwan, 1999, pp. 466-477.

/29/ Ellingwood B R, LRFD: implementing structural reliability in professional practice, *Engineering Structures*, 22, 2000, pp. 106-115.

/30/ Enright M P and Frangopol D M. Time-variant system reliability prediction, in *Probabilistic Mechanics and Structural Reliability*, A. Kareem, A. Haldar, B.F. Spencer and E.A. Johnson, eds., ASCE, PMC 2000-053, 2000, 6p on CD-ROM.

/31/ Fisher J W, Considerations underlying life cycle engineering of infrastructure," in *Workshop on integrated research for civil infrastructure systems,* R. Zimmerman and R. Sparrow, eds., New York University, New York, 1997, F-19&20.

/32/ Ang A H-S and De Leon D, Target reliability for structural design based on minimum expected life-cycle cost, in *Reliability and Optimization of*

Structural Systems, D.M. Frangopol, R.B. Corotis, and R. Rackwitz, eds., Pergamon, 1997, pp. 71-83.

/33/ Ang A H-S, Frangopol D M, Ciampoli M, Das P C and Kanda J, Life-cycle cost evaluation and target reliability for design, in *Structural Safety and Reliability*, N. Shiraishi, M. Shinozuka, and Y.K. Wen, eds., Balkema, Rotterdam, 1998, pp. 77-78.

/34/ Frangopol D M, Integration of life-cycle cost analysis with safety analysis, BASSAR, *Structural Engineering International*, Journal of IABSE, 10, 2000, p. 278.

/35/ Frangopol D M, Life-cycle cost optimization of deteriorating systems, Keynote Paper in *Proc. of the Fourth Japan Conference on Structural Safety and Reliability*, JCOSSAR 2000, Tokyo, Japan, Vol. 4, 2000, pp. 1-8.

/36/ Tilly G P, Principles of whole life costing, in *Safety of Bridges*, P.C. Das, ed., Thomas Telford, 1997, pp. 138-144.

/37/ Van Noortwijk J M, Optimal replacement decisions for structures under stochastic deterioration, in *Reliability and Optimization of Structural Systems*, A.S. Nowak, ed., University of Michigan, Chelsea, Michigan, 1998, pp. 273-280.

/38/ Frangopol D M, Lin K-Y and Estes A C, Life-cycle cost design of deteriorating structures, *Journal of Structural Engineering*, ASCE, 123 (10), 1997, pp. 1390-1401.

/39/ Mori Y and Ellingwood B R, Maintaining reliability of concrete structures-II: Optimum inspection/repair, *Journal of Structural Engineering*, ASCE, 120 (3), 1994, pp. 846-862.

/40/ Ellis J H, Jiang M and Corotis R B, Inspection, maintenance and repair with partial observability, *Journal of Infrastructure Systems*, ASCE, **1** (2), 1995, pp. 92-99.

/41/ Wen Y K and Kang Y J, Design based on minimum expected life-cycle cost, in *Advances in Structural Optimization,* D.M. Frangopol and F.Y. Cheng, eds., ASCE, New York, 1997, pp. 192-203

/42/ Ang A H-S, Lee J-C and Pires J A, Cost-effectiveness evaluation of design criteria, in *Optimal Performance of Civil Infrastructure Systems*, D.M. Frangopol, ed., ASCE, Reston, Virginia, 1998, pp. 1-16.

/43/ Wallbank E J, Tailor P and Vassie P R, Strategic Planning of Future Maintenance Needs, in *Management of Highway Structures*, P.C. Das, ed., Thomas Telford, London, 1999, pp. 163-172.

/44/ Ellingwood B R, Optimum policies for reliability assurance of aging concrete structures, in *Optimal Performance of Civil Infrastructure Systems*, D.M. Frangopol, ed., ASCE, Reston, Virginia, 1998, pp. 88-97.

/45/ Frangopol D M, Life–cycle cost analysis for bridges, Chapter 9 in *Bridge Safety and Reliability,* D.M. Frangopol, ed., ASCE, Reston, Virginia, 1999, pp. 210-236.

/46/ Kong J S and Frangopol D M, Life-cycle reliability-based maintenance cost optimization of deteriorating structures with emphasis on bridges, *Journal of Structural Engineering*, ASCE, 129(6), 2003, pp. 818-828.

/47/ Kong J S and Frangopol D M, Evaluation of expected life-cycle maintenance cost of deteriorating structures, *Journal of Structural Engineering*, ASCE, 129(5), 2003, pp. 682-691.

/48/ Frangopol D M and Maute K., Life-cycle reliability-based optimization of civil and aerospace structures, *Computers & Structures*, Pergamon, 81(7), 2003, pp. 397-410.

/49/ Estes A C and Frangopol, Life-cycle evaluation and condition assessment of structures, Chapter in *Structural Engineering Handbook*, Second Edition, CRC Press, 2004 (in press).

/50/ Yang S-I, Frangopol D M and Neves L C, Service life prediction of structural systems using lifetime functions with emphasis on bridges, *Reliability Engineering & System Safety*, Elsevier, 2004 (in press).

Chapter 7

Reliability-based Service Life Prediction and Durability in Structural Safety Assessment

By:
Yasuhiro Mori, Nagoya University, Japan

7.1 INTRODUCTION

In most applications of structural reliability analysis and, in particular, in the development of reliability based codes, the resistance of a structure has been assumed to be a time-invariant random variable. Only temporal variations in loads have been taken into account. There are, however, many environmental stressors and repetitive loads that can cause deleterious changes in strength and stiffness beyond the baseline conditions assumed for design (Clifton and Knab, 1989). Some of these effects are benign; others may cause strength to degrade over extended periods of time. In reinforced concrete structures, for example, carbon dioxide penetrates the concrete cover on the reinforcement and may lead to the initiation of corrosion, resulting in concrete cracking and a reduction of structural capacity as the corrosion progresses (Tuutti, 1982). Corrosion of a steel component causes loss of material leading to smaller net section and accordingly increases the stress level for a given load. These changes impact the safety and serviceability of a structure, and the issue of durability arises.

One cannot deal with durability issues rationally without introducing the notion of service life (Somerville, 1986). The service life of components with a structural function is defined as the period during which the structure is able to withstand all loads without loss of function. A design service life (or maintenance interval) is the period during which the probability of the structure performing its intended functions is acceptable. For example, carbonation of concrete, in itself, does not cause any structurally significant change in the strength of a reinforced concrete component; the structural impact comes only after the carbonation front reaches the level of the steel reinforcement, leading to surface depassivation and the possible initiation of corrosion. The limit state probability must be determined with respect to a service life in order to be useful as a decision variable.

This chapter describes the methodologies to evaluate structure reliability taking into account the stochastic nature of future load, randomness in strength and in degradation resulting from environmental stressors and repetitive loads. The effect of in-service inspection and repair in assuring the safety and serviceability of a structure is considered taking the imperfect nature of non-destructive evaluation (NDE) technique into account. Optimum in-service inspection and repair strategies can be determined within this framework to meet performance goals stated in probabilistic terms.

Durability is an important issue especially for a tall building. Because of technical difficulties in demolishing such a huge structure, necessity of large investment, it is not easy to re-construct the building. Furthermore, because of environmental problems such as the disposal of large amount of waste and discharge of large amount of CO_2 during re-construction, it is expected that the service life of a tall building would be much longer than that of an ordinary building. Then the deterioration of the structure, which could be neglected for an ordinary building with a short service life, as well as the inspection and maintenance to mitigate its impact on the structural safety, should be taken into account in the assessment of a tall building.

7.2. DEGRADATION MECHANISMS AND STRUCTURAL RESISTANCE

The service life of structural components and systems is dependent on how the material properties, and thus structural resistance, change over time. Such changes are brought about by naturally occurring chemical and physical processes, external environmental stressors, environmental or accidental events, wear during service operation, and improper use and maintenance. In this section, material properties, environmental factors and degradation mechanisms that are most important in determining the service life of reinforced concrete structures and steel structures are discussed.

7.2.1 CONCRETE STRUCTURES

The singularly most important factor affecting concrete quality is its permeability, which controls the rate of ingress of aggressive substances. The permeability of hardened concrete can be reduced by lowering the water/cement ratio, using pozzolanic materials and proper placement and curing. High quality concrete with low permeability may maintain its integrity and have a service life of several hundred years if the conditions to which it is exposed are not severe (Somerville, 1986).

Properly placed concrete increases in strength due to continued hydration well beyond its specified compression strength at 28 or 90-days, f_c, used as a basis for design (Washa and Wendt, 1975). It is not uncommon for concrete strength to increase by a factor of 2 or more over 50 years if it is protected from aggressive environments. Experiments conducted over a period of 50 years (Washa and Wendt, 1975; Washa, et al, 1989) show that the increase in compressive strength is roughly proportional to the logarithm of the age during the first 10 to 50 years and the strength changes very little thereafter. Thus, $f_c(t)$ can be modeled as,

$$f_c(t)/f_{c_{28}} = \begin{cases} \alpha + \beta \ln t & t < t_M \\ \alpha + \beta \ln t_M & t \geq t_M \end{cases} \tag{7.1}$$

in which $f_{c_{28}}$ = 28-day compressive strength, t = time in days, and age to maturity t_M depends on the chemical composition of concrete. For concrete made with relatively low C_2S, which contributes the long-term strength of concrete, $t_M \approx 10$ years, and $f_c(t_M)/f_{c_{28}} = 1.67$ yielding $\alpha = 0.541$ and $\beta = 0.138$. The coefficient of variation (c.o.v.) for the 50-year compressive strength for a given set of variables is low, varying between 5% and 10 %.

Concrete components, however, may degrade over an extended period of time due to aggressive environmental stressors and/or loads. For concrete strength, the most significant stressors are freeze-thaw cycling, sulfate attack, alkali-silicate reactions within the concrete, and fatigue from repetitive loads. For deformed bar reinforcement, the possibility of corrosion is the most important; for prestressing tendons, detensioning due to tendon relaxation, anchorage failure or creep in the concrete must be considered in addition to the factors that affect deformed bars.

In general, the depth of attack or deterioration, $X(t)$, can be modeled by,

$$X(t) = \begin{cases} 0 & ; t < T_I \\ C(t - T_I)^{\alpha} & ; t \geq T_I \end{cases} \tag{7.2}$$

in which t = elapsed time, T_I = time required to activate the deterioration process, C = rate parameter dependent on the aggressive stressor and concrete mix, and α = time-order parameter dependent on the basic nature of the attack. The parameters C and α must be determined from experimental data. Fig.7.1 illustrates conceptually the general shapes expected for different mechanisms (Clifton and Knab, 1989). For processes that are essentially diffusion-controlled α =1/2; for other degradation mechanisms, α may be greater than unity.

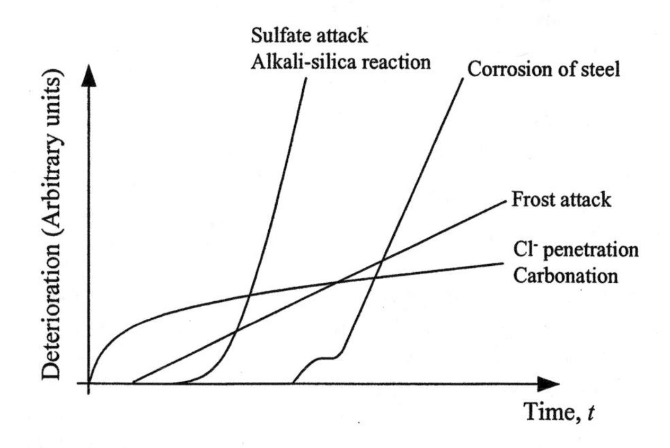

Figure 7.1: Material Degradation (Clifton and Knab, 1989)

Corrosion of reinforcement is one of the most damaging mechanisms affecting the strength of reinforced or pre-stressed concrete structures over time. A review of data on uniform corrosion for carbon steels indicates that the rate parameter, C, typically ranges from about 50 - 125 for $X(t)$ measured in μm and in years; parameter α typically is between 1/2 and 1. The variability in C is quite large, its coefficient of variation (c.o.v.) being typically close to about 0.5. In comparison, α can be assumed to be deterministic.

The time to initiate corrosion, T_I, clearly depends on the amount of concrete cover, permeability, and the presence of cracks on the concrete surface. Penetration of the concrete by atmospheric carbon dioxide reduces the natural alkalinity of the concrete. Carbonation is basically a diffusion-controlled process. The depth of carbonation, $X(t)$, is given as,

$$X(t) = K\sqrt{t} \tag{7.3}$$

in which K = experimental constant. If concrete is sheltered from rain, K is estimated by (Tuutti, 1982)

$$K = 26(W/C - 0.3)^2 + 1.6 \qquad (\text{mm/year}^{1/2}) \tag{7.4}$$

where W/C is the water-cement ratio of the concrete, by weight.
Cracking in concrete components and structures may have little effect on their strength. However, the carbon dioxide in the air penetrates the concrete more easily at a crack (Vesikari, 1988), and the initiation period of carbonation decreases dramatically. In this case, the depth of carbonation at a crack is obtained from,

$$X(t) = 50\sqrt{w}\ \sqrt[4]{t} \quad (\text{mm}) \tag{7.5}$$

where w is the crack width (mm).

Active corrosion begins once the level of carbonation or aggressive ion penetration has reached the reinforcement. In previous studies, T_I has been assumed to be log normally distributed with a c.o.v. of 0.20 (Vesikari, 1988).

Structural resistance, $R(t)$, can be related to $X(t)$ for a given behavioral limit state of interest, such as flexure, shear, or compression. The relation between $R(t)$ and $X(t)$ may be nonlinear, depending on the nature of the attack and the mechanics of the limit state. For example, the section loss of a reinforcing bar due to corrosion would cause the decrease of its tensile strength according to,

$$R(t) = R_0 \cdot \left[1 - \frac{2 \cdot C \cdot (t - T_I)^{\alpha}}{D}\right]^2 \tag{7.6}$$

in which R_0 = tensile strength of reinforcing bar in the original state. The flexural strength of a beam or a slab would decrease accordingly.

7.2.2 STEEL STRUCTURES

Corrosion and fatigue are two major deterioration mechanisms of steel structures. Corrosion causes loss of material leading to smaller net section and accordingly increases the stress level for a given load. The geometric parameters, such as moment of inertia and radius of gyration, decrease due to the reduction of section area. Buckling capacity of members can be critically affected by the reduction in metal thickness.

The depth of corrosion penetration also follows Eq. (7.2) (Kemp, 1987). The rate of corrosion depends on the amount of moisture and ambient temperature. The parameters C and α in Eq. (7.2) can be determined from experimental data. Those for carbon and weathering steel in exposed environments are summarized in Table 7.1 for $X(t)$ measured in μm and in years (Albrecht and Naeemi, 1984). Some of the results could be applied to the components in building structures.

Table 7.1: Average Value for Corrosion Parameters

Environment	Carbon steel		Weathering steel	
	C	α	C	α
Rural	34.0	0.65	33.3	0.50
Urban	80.2	0.59	50.7	0.57
Marine	70.6	0.79	40.2	0.56

A steel component under cyclic load would deteriorate due to the propagation of micro cracks, called fatigue cracks. Components in a tall building would be prone to fatigue because of high intensity of repetitive wind loads. In practice, fatigue life prediction has often been carried out deterministically based on the endurance of a component (that is the number of cycles to cause the component to fracture at the peak of the cyclic load). However, such endurance does not provide any information about the variation in the residual strength of a component with cyclic loading (Oehlers, et al, 1995).

As fatigue behavior is highly variable in nature, it should be treated in a probabilistic sense. Applying stochastic fracture mechanics, various probabilistic models have been proposed to describe the stochastic characteristics of crack growth under random loadings (Oswald and Schueller, 1984; Oehlers, et al, 1995; Zheng and Ellingwood, 1999). Taking the randomness in load occurrence and intensity into account, the probabilistic characteristics of crack length can be estimated as a function of time.

7.3. TIME-DEPENDENT RELIABILITY ANALYSIS

7.3.1 STOCHASTIC LOAD MODELS

Events giving rise to significant structural loads occur randomly in time and are random in intensity. When viewed on a time scale of 50 years or more, the duration of design-basis events generally is very short, and thus such events occupy only a small fraction of the total life of a component. With these assumptions, a structural load that varies in time can be modeled as a sequence of randomly occurring pulses with random intensity, S_j, and duration, τ, as illustrated in Fig.7.2.

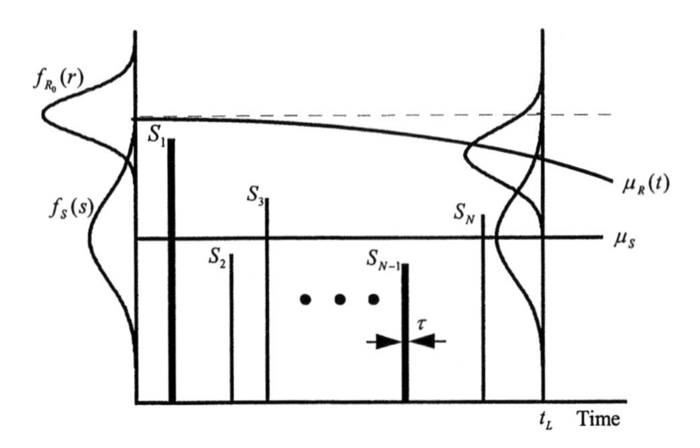

Figure 7.2: Schematic Representation of Load Process and Degradation of Resistance

Based on the above description of design events, the occurrence in time of a particular event may be described by a Poisson point process. With the Poisson model, the probability that $N(t) = n$ load events occur within the interval $(0, t]$ is,

$$P[N(t) = n] = \frac{(\lambda t)^n \cdot \exp(-\lambda t)}{n!}; \quad n = 0, 1, 2, ... \tag{7.7}$$

in which $P[\bullet]$ = probability of event in brackets, and λ = mean rate of occurrence of the events. The sequence $S_j, j = 1, 2, ..., N(t)$, are assumed to be identically distributed and statistically independent random variables described with cumulative distribution function (cdf) $F_S(s)$.

7.3.2 DEGRADATION FUNCTION

The structural capacity, $R(t)$, at time t degrades according to,

$$R(t) = R_0 \cdot G(t) \tag{7.8}$$

in which R_0 = component capacity in the under graded (original) state and $G(t)$ = time-dependent degradation function defining the fraction of initial strength remaining at time t. For some of the degradation factors, the initial strength and degradation function are separable, e.g. deterioration due to corrosion or sulfate attack does not depend on the initial strength of the reinforcement or concrete. For the others, degradation function depends on the initial strength. Function $G(t)$ may not be monotonic because of the presence of several aging mechanisms in combination, e.g., some aging mechanisms may degrade concrete strength while others may cause the strength to increase. Due to uncertainty in the degradation mechanisms and the lack of experimental data, the function $G(t)$ is generally a stochastic process.

7.3.3 RELIABILITY OF COMPONENT DEGRADING DUE TO AGGRESSIVE ENVIRONMENT

To illustrate the basic concepts of time-dependent reliability analysis, consider a structural component subjected to the sequence of discrete stochastic load events illustrated in Fig.7.2. Assume that the component degrades with time described by a degradation function, $G_a(t)$, due to aggressive environment and that $G_a(t)$ is described by an independent of the load history. Since it has been found that the variability in $G_a(t)$ is of minor importance when compared to mean degradation and load process characteristics, it is assumed that $G_a(t)$ is deterministic and equal to mean $E[G_a(t)] = g_a(t)$. Then the time-dependent reliability function (probability of survival in interval $(0,t_L])$ is represented as follows (Mori and Ellingwood, 1993b):

$$L(t_L) = \int_0^\infty \exp\left[-\lambda \cdot t_L \cdot \left[1 - \frac{1}{t_L}\int_0^{t_L} F_S\{r \cdot g_a(t)\}dt\right]\right] f_{R_0}(r)dr \quad (7.9)$$

in which λ = mean occurrence rate of load events and $f_{R_0}(r)$ = probability density function (pdf) of R_0, expressed in units that are dimensionally consistent with S. If a component is subjected to two statistically independent load processes, S_1 and S_2 but only S_1 varies in time, then from Eq. (7.9) and the theorem of total probability the reliability function becomes

$$L(t_L) = \int_0^\infty \int_0^\infty \exp\left[-\lambda_{S_1} \cdot t_L \cdot \left[1 - \frac{1}{t_L}\int_0^{t_L} F_{S_1}\{r \cdot g_a(t) - s_2\}dt\right]\right] f_{S_2}(s_2) f_{R_0}(r) ds_2 dr \quad (7.10)$$

in which λ_{S_1} = mean occurrence rate of S_1, F_{S_1} = cdf of S_1, and f_{S_2} = pdf of S_2.

The limit state probability, or probability of failure, is given by,

$$F(t_L) = 1 - L(t_L) \quad (7.11)$$

The hazard function, $h(t)$, defined as the probability of failure within time interval $(t, t+dt)$ given that the component has survived up to time t, can be expressed as,

$$h(t) = -\frac{d}{dt}\ln\{L(t)\} \quad (7.12)$$

When failure is due to structural aging, $h(t)$ increases with time.

7.3.4 RELIABILITY OF COMPONENT DEGRADING DUE TO REPETITIVE LOADS

The strength of a structure or a component may also degrade due to external loading according to (Kameda and Koike, 1975),

$$R_i = R_{i-1} \cdot \phi(S_i) \tag{7.13}$$

in which R_i = residual strength after the i-th load event and $\phi(s)$ (≤ 1) = degradation factor due to a load event with intensity s. R_i can be expressed in terms of the initial strength, R_0, as,

$$R_i = R_0 \cdot \prod_{j=1}^{i} \phi(S_j) \tag{7.14}$$

$\phi(s)$ may depend on the load intensity and structural resistance.

If the degradation function of a component due to repetitive loads can be approximately expressed as a function of time and independent of random load intensities, the degradation function of a component due to aging and repetitive loads, $G_\ell(t)$, also becomes independent of random load intensities. Moreover, if the variability in $G_\ell(t)$ is small, the reliability function of a component degrading due to aging and repetitive loads can be estimated by Eq.(7.9) replacing $g_a(t)$ with $g_\ell(t)(=E[G_\ell(t)])$ (Mori and Ellingwood, 1995).

$$\begin{aligned} g_\ell(t) &= \exp\left[-\lambda \cdot t \cdot \left\{1 - \int_0^\infty \phi(s) f_S(s) ds\right\}\right] \\ &= \exp[-\lambda \cdot t \cdot \{1 - E[\phi(S)]\}] \end{aligned} \tag{7.15}$$

If the variability in the effect of degradation due to repetitive loads is small, $g_\ell(t)$ can be approximately evaluated by,

$$g_\ell(t) = \exp[-\lambda \cdot t \cdot \{1 - \phi(\mu_S)\}] \tag{7.16}$$

When a structure or a component degrades due to both aging and repetitive loads, the degradation function, $G(t)$, can be expressed in terms of each degradation function. For example, if the material strength degrades only due to repetitive loads while section area decreases only due to aging and if these degradation mechanisms are independent of each other, $G(t)$ can be expressed by the product of $G_a(t)$ and $G_\ell(t)$ as,

$$G(t) = G_a(t) \cdot G_\ell(t) \tag{7.17}$$

If there is a interaction between these two degradation mechanisms, the degradation model would be more complicated.

7.3.5 STRUCTURAL SYSTEM

A structural system composed of beams, columns, slabs, and walls can be modeled by a combination of two fundamental subsystems: series systems and parallel systems (Melchers, 1987). A series system fails if any of its components fails, while a strictly parallel system fails only if all its components fail. Because structural redundancy is more closely modeled by a parallel system, the reliability of a structure modeled as a series system of components provides a conservative lower bound estimate of the actual system reliability. The time-dependent reliability function for a system, $L_S(t_L)$, becomes (Mori and Ellingwood, 1993b),

$$L_S(t_L) = \underbrace{\int_0^\infty \cdots \int_0^\infty}_{m\text{-fold}} \exp\left[-\lambda \cdot t_L \cdot \left\{1 - \frac{1}{t_L}\int_0^{t_L} F_S\left(\min_{i=1}^{m} \frac{r_i \cdot g_i(t)}{c_i}\right) dt\right\}\right] \cdot f_{\underline{R}}(\underline{r}) d\underline{r} \tag{7.18}$$

in which R_i = the initial strength of component i, $\underline{R} = \{R_1, \ldots, R_m\}$, $f_{\underline{R}}(\underline{r})$ = the joint probability density function of $\underline{R}$, $g_i(t)$ = the degradation function of component i, and $c_i \cdot s$ is the structural action (e.g., moment, axial force, shear, etc.) induced on component i by a load with s acting on the structure.

Eq. (7.18) cannot be integrated in closed form because of the (m+1)-fold integration. The analysis of even a 2-component system requires a relatively long computation time. An importance sampling technique (e.g., Melchers, 1990; Mori, 1993a) can be applied to determine the time-dependent system reliability.

7.3.6 ILLUSTRATION OF TIME-DEPENDENT RELIABILITY

1. Component Reliability

The effect of degradation in component strength on the component reliability function is illustrated using several simple parametric representations of time-dependent strength to demonstrate those parameters that appear to have a particularly significant impact on reliability.

Table 7.2: Parameters of Load Process and Resistance

	mean*	c.o.v.	pdf	λ (yr^{-1})	τ
Dead Load	$1.00D_n$	0.10	Normal	–	–
Sustained Live Load	$0.30L_n$	0.40	Lognormal	–	–
Wind Load	$0.60W_n$	0.45	Type I	1	6 hr.
Resistance	$1.10R_n$	0.15	Lognormal	–	–

*D_n, L_n, and W_n are nominal loads and R_n is nominal strength

Table 7.3: Degradation Model

Shape of the Degradation Function	Degradation Rate $g(80)$	Corresponding Degradation Mechanism
Linear: $g(t) = 1 - at$	0.7, 0.8, 0.9	Corrosion
Parabolic: $g(t) = 1 - at^2$	0.8, 0.9	Sulfate Attack
Square root: $g(t) = 1 - a\sqrt{t}$	0.8, 0.9	Diffusion Controlled Degradation

The combination of time varying wind load, W, and time-invariant dead load, D

and sustained live load L, is considered. The probabilistic models of load intensity and resistance used in this parametric sensitivity study are summarized in Table 7.2. These statistics are selected for illustrative purposes. In each analysis, reliability is evaluated for a service period up to 120 years.

The degradation models summarized in Table 7.3 were selected to examine the sensitivity of the time-dependent reliability to the type of degradation. The linear and parabolic functions model, in an approximate manner, degradation in strength due to corrosion of reinforcement and sulfate attack, respectively. The square root function represents a degradation process that is diffusion-controlled and where the degradation rate decreases in time. The degradation rate parameter is given with reference to the residual strength at 80 years; for example, a linear model with $g(80)$ = 0.8 (based on remaining reinforcing bar area) might correspond to the case where the initial diameter of the reinforcement is 16 mm and the corrosion rate is constant and approximately equal to 0.04mm/year.

The mean value of initial strength is determined referring to the design requirement described in the guideline of limit state design (AIJ, 1998),

$$0.9R_n = 1.0D_n + 0.6L_n + 1.6W_n \tag{7.19}$$

in which R_n = nominal capacity, and D_n, L_n, and W_n = nominal dead load, nominal sustained live load, and nominal wind load, respectively. Assuming that $L_n = D_n$ and $W_n = 2D_n$, Eq.(22) provides the central factor of safety, $\mu_R/(\mu_D + \mu_L + \mu_W)$ = 2.40, and the reliability index, β = 1.82 for the service life of 50 years. Taking longer service life and deterioration into account, the mean value of the initial strength is increased by 15% for a component considered in the numerical examples. The reliability index of the component for the service life of 100 years is 2.20, if the component does not degrade.

In the course of the analysis, the effect of variability in the combination of dead load and sustained live load considered in the illustrations was found to be negligible compared with the effect of other factors, and thus they are treated as a deterministic variables and equal to its mean values in the followings.

The effect of the overall degradation in 80 years on the failure probability is presented in Fig.7.3, using the linear degradation model. When the residual strength at 80 years decreases by 0.1, the failure probability increases by the factor of 2 - It exceeds an acceptable level (e.g., β = 1.82 for the service life of 100 years) if $g(80) \le 0.8$. The effect of the shape of the degradation function on the limit state probability, $F(t)$, is presented in Figs. 7.4 for (a) $g(80) = 0.9$ and (b) $g(80) = 0.8$. Up to 100 years, the failure probability associated with the square root model is the highest, followed by $F(t)$ for the linear and parabolic models. However, after 100 years, the failure probabilities associated with the parabolic model increase rapidly. These results can be explained by the fact that, for the same fraction, $g(80)$, of residual strength at 80 years, the degradation in component strength using the square root model occurs mainly in the early stages of the reference period while the component strength using the parabolic model degrades more rapidly in the late stages. The shape of the degradation functions has a greater influence as $g(80)$ decreases. For lower residual strength at time t, the failure probability is sensitive to small changes in the strength; this effect accumulates with time. Figs.7.3 and 7.4 demonstrate the importance of the quantitative evaluation of the parameters that define the strength degradation functions.

Figure 7.3: Dependence of Single Component Failure Probability on Degradation Rate

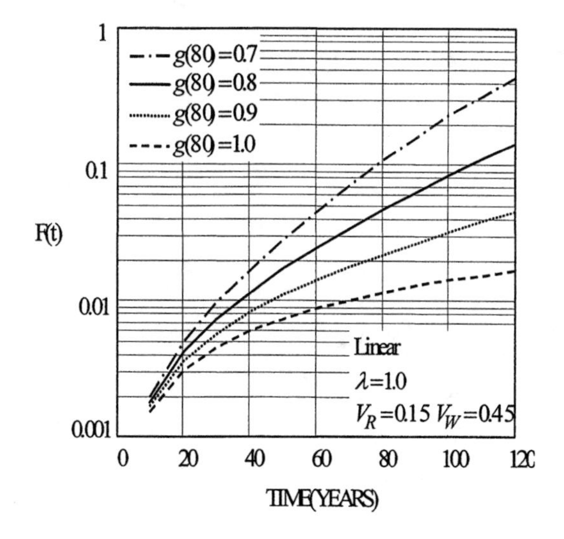

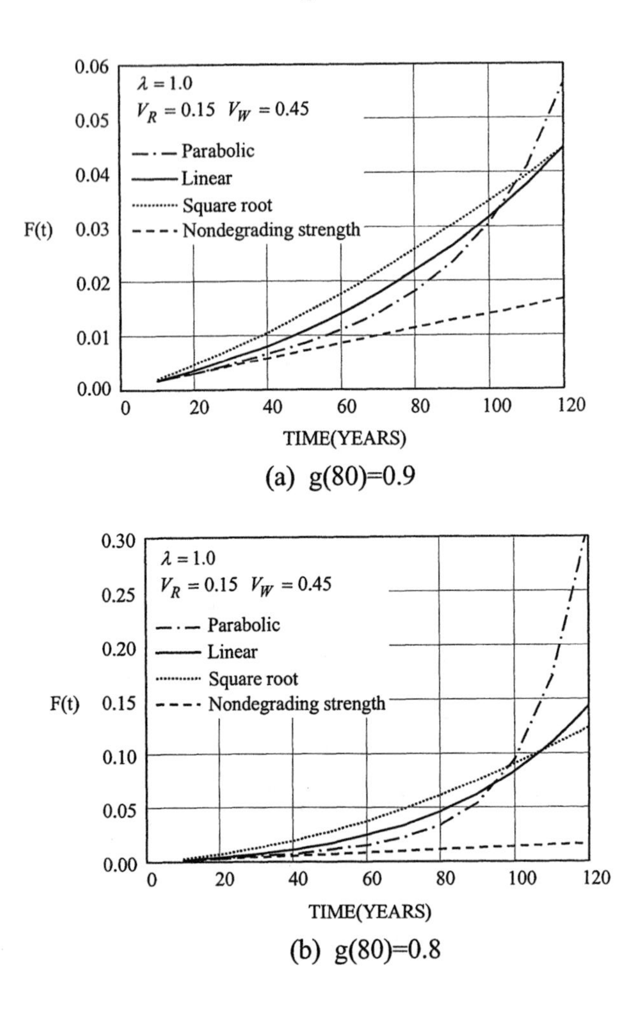

Figure 7.4: Dependence of Single Component Failure Probability on Degradation Model

Linearly increasing hazard functions have been suggested by some investigators as possible aging models (Vesely, 1987). The hazard functions, $h(t)$, associated with the strength degradation models presented in Figs.7.4(a) and 7.4(b) are illustrated in Figs.7.5(a) and 7.5(b). The hazard function associated with the square root model is nearly linear in time for $g(80) = 0.9$. However, $h(t)$ is nonlinear for the linear and parabolic degradation models, and its curvature increases as $g(80)$ decreases. Therefore, linear hazard functions might be reasonable if the structural degradation can be described by a diffusion process with a small rate of degradation; such models, however, generally may not be flexible enough to analyze deterioration in structural components.

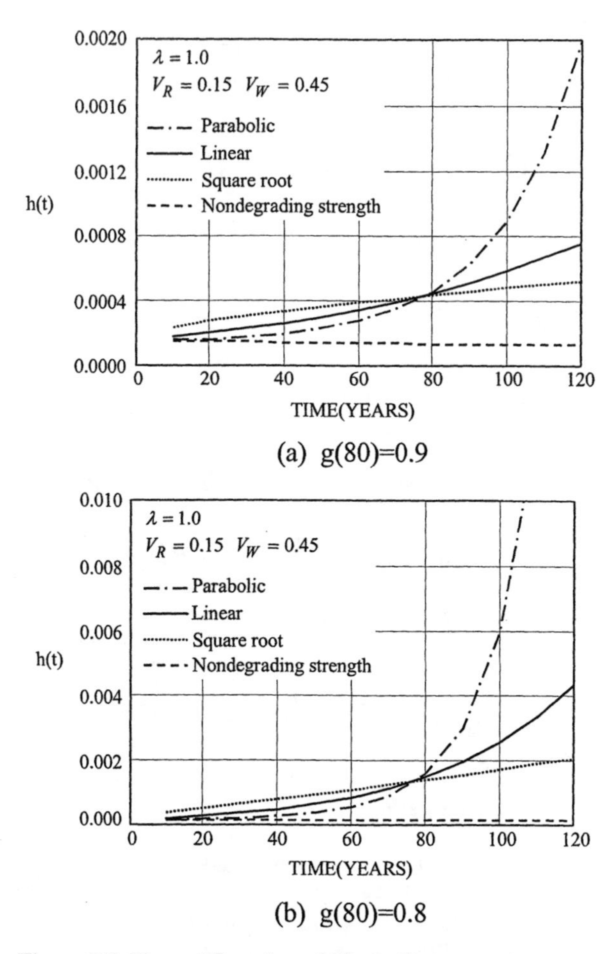

Figure 7.5: Hazard Function of Single Component

In the illustrations in Figs.7.3 through 7.5, only the active stage of component strength degradation is considered. In Fig.7.6, the effect of the time to initiate degradation, T_I, is illustrated using linear strength degradation expressed as,

$$g(t;t_I) = \begin{cases} 1 & ;t < t_I (23) \\ 1 - a(t - t_I) & ;t \geq t_I \end{cases} \tag{7.20}$$

in which the degradation rate, a, is given by $g(80;T_I = 0) = 0.8$. It is assumed that the induction time is deterministic and equals to $t_I = 10, 20,$ and 30 years. It can be seen in Fig.7.6 that ignoring the initiation phase may lead to an overly pessimistic appraisal of reliability. As illustrated in Figs.7.4, the failure probabilities computed using both linear and parabolic strength degradation models increase mainly in later stages of exposure. Since whether or not this late stage is included during the service life is dependent upon the initiation time, the appropriate evaluation of induction time is important to the reliability assessment.

The variability in T_I, on the other hand, has a secondary effect on the reliability assessment (Enright and Frangopol, 1998). This can be observed in Fig.7.6 where the difference of induction time in the order of 10 years yields a small difference in failure probability.

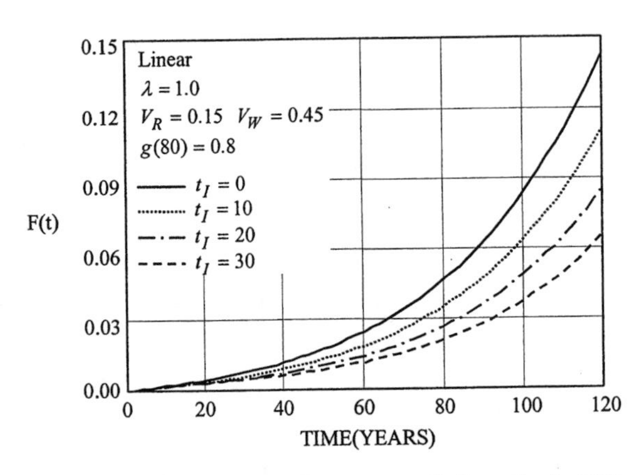

Figure 7.6: Dependence of Single Component Failure Probability on Initiation Time

7.3.7 System Reliability

The effect of degradation in component strength on the system reliability is illustrated for the load combination $D+L+W$ using the same load and resistance parameters as in the illustrations of component reliability and the linear degradation model. Without loss of generality, the c_i's in Eq.(7.18) are assumed to be unity for all components. In each analysis, a five-component system is considered. The degradation rate equals 0.8 at 80 years for all components unless otherwise specified.

The strengths of the components in a system are correlated because of common construction materials, fabrication, and construction practices. The sensitivity of the system reliability to the stochastic dependence among component strengths is investigated by assuming that the strengths of the components are identically distributed and equally correlated pair-wise with correlation coefficient ρ. The failure probability increases as the correlation coefficient decreases, as shown in Fig.7.7. However, the effect of ρ is relatively small, particularly when $\rho < 0.8$, and decreases as the coefficient of variation of load intensity increases. The reduction in the sensitivity of the reliability function to ρ is due to the dependence among component failure events that arises from common random loads with relatively large variability.

It is unreasonable to expect that perfect correlation exists among the strengths of components. In light of the relative insensitivity to ρ, it seems appropriate to model systems such as those considered herein as being composed of components with independent strengths, particularly when the coefficient of variation of load intensity is large relative to that of resistance. Although this model is conservative, the reliability is substantially less affected by this assumption than it is by the choice of strength degradation model.

In general, a system consists of components with different statistical characteristics of strength and because of, some components affect the system failure probability more than others. Such critical components are a result of basically two sources: (a) Differences in initial safety factors or statistics of initial strength, and (b) Differences in degradation mechanism and associated degradation rate and time to initiate degradation. The combination of these cases also may cause a change in which component is critical at a particular time.

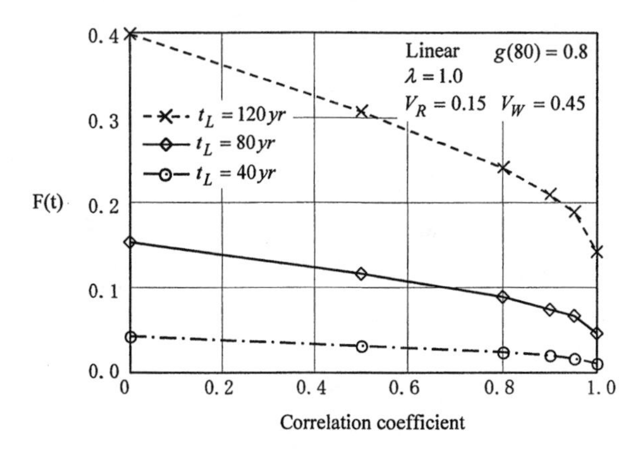

Figure 7.7: Dependence of System Failure Probability on ρ

The effect of the statistics of initial strength of components on system reliability is shown in Fig.7.8. The mean initial strengths of all components except one (the critical component) are increased by 5%, 10%, or 20%. It is assumed that the initial strengths of the components are statistically independent of one another. As the mean strength of non-critical components increases, the failure probability of the system approaches that of a single critical component. If the mean strengths of non-critical components are increased by more than 20%, the contribution of these components to system reliability is negligible and the size of the system for reliability analysis purposes effectively can be reduced by treating those components as if their strengths are deterministic.

The effect of component degradation rates on system reliability are shown in Fig.7.9. The degradation rates are set so that $g(80) = 0.8$ for one component (critical component) while $g(80) = 0.9$, 0.95 or 0.98 for the other components (non-critical components). The initial strengths of all the components are assumed to be identically and log normally distributed and statistically independent of one another. These system failure probabilities are bounded by the failure probability of the single critical component in the system and the failure probability of a system consisting of only critical components. At the early stage of the service life when all components remain close to their initial strength, the system failure probability is close to that of the system in which all components degrade equally and the non-critical components still contribute to the system failure. Later in the service life, however, the contribution of non-critical components decreases and the system failure probability approaches that of the single critical component. After about 80 years, the effect of non-critical components is negligible and the system reliability can be determined approximately as that of the single critical component. Therefore, identification of critical components is important in a time-dependent reliability analysis of a complex system because it enables the analyst to keep the system analysis to a manageable size.

Further sensitivity analysis to various parameters can be found elsewhere (Enright and Frangopol, 1998).

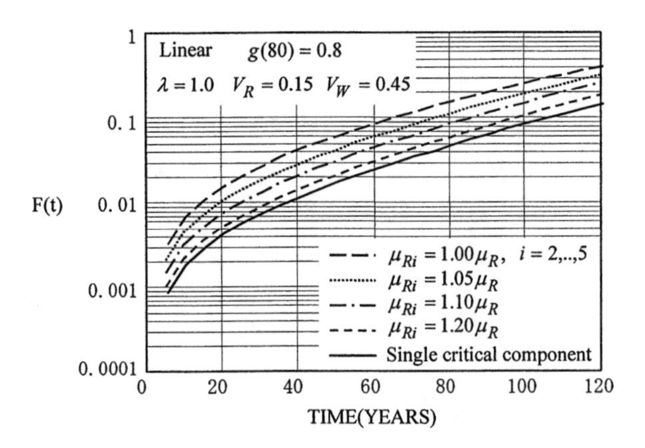

Figure 7.8: Dependence of System Failure Probability on Initial Strength

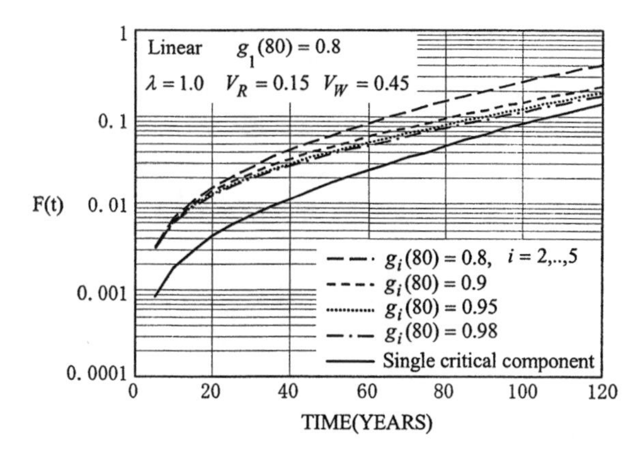

Figure 7.9: Dependence of System Failure Probability on Degradation Rate

7.4. IN-SERVICE INSPECTION AND REPAIR

7.4.1 IMPACT OF IN-SERVICE INSPECTION AND REPAIR

The analysis of reliability of components and structures as functions of time enables the analyst to determine the service period beyond which the desired reliability of the structure cannot be ensured. To extend the service period, the structure may have to be inspected and maintained. The intervals and extent of inspection and maintenance that are required to ensure the performance during its intended service life can be determined using the time-dependent reliability analysis presented in the previous section. Some studies have been done to determine optimal inspection/maintenance strategies for steel structures subjected to fatigue, assuming that a component is replaced if the intensity of detected damage exceeds a critical value (e.g., Thoft-Christensen and Sorensen, 1987). However, rather than replacing a component, one might only repair detected damages. In this case, the strength of the component may not be fully restored to its original level because of the effect of damage overlooked at an inspection.

An ideal non-destructive evaluation (NDE) technology would have a high probability of locating flaws and, at the same time, a low probability of false indications that might lead to unnecessary and costly repair. No NDE can detect small defects with certainty. The imperfect nature of NDE methods is random and can be described in statistical terms. Flaw detectability can be described by a function in the form of a cumulative distribution, $d(x)$, which simply expresses the probability that an NDE device will be able to detect a flaw with size x. Examples of $d(x)$ are illustrated in Fig.7.10. Such a relation exists, at least conceptually, for each in-service inspection technology. Practically, $d(x)$ often has a large dispersion for field inspection and sufficient data seldom exist to define $d(x)$. Sensitivity studies incorporating different $d(x)$ in time-dependent reliability analyses indicate that the anchor point, $\mu_{X_{th}}$, of $d(x)$, is more important than the shape of $d(x)$ (Mori and Ellingwood, 1993c). Thus, $d(x)$ might be approximated for a practical purpose by a step function as illustrated in Fig.7.10.

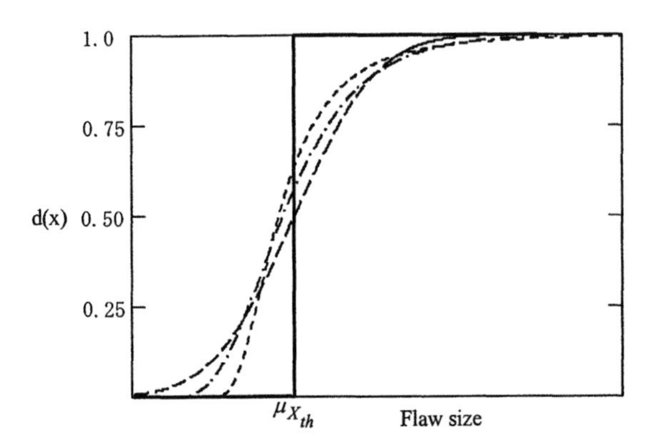

Figure 7.10: Probability of detecting flaw of size x or larger

In a general sense, when an inspection is performed, we learn something about the condition of the structure that may allow us to update the probabilistic strength models and degradation models used in the time-dependent reliability analysis (e.g., Ciampoli, 1989; Mori and Ellingwood, 1994a; Enright and Frangopol, 1999). The time-dependent reliability analysis is re-initialized using the updated $f_{R_0}(r)$ and $g(t)$ in Eq.(7.10).

To illustrate the impact of in-service inspection and repair on the reliability of structural component, two simple strategies are considered: (1) infrequent inspection and maintenance carried out at 40 and 80 years, with a NDE technique capable of detecting only defects causing an 1% reduction in strength, and (2) frequent inspection and maintenance performed at the age of 30 years and every 10 years thereafter with a NDE technique capable of detecting only defects causing an 5% reduction in strength. It is assumed that all detected flaws are repaired completely. Thus, the strength of a component immediately after repair equal to the 99% of the original state for strategy (1), while the strength is equal to the 95% of the original state for strategy (2). Strength degradation is assumed to occur linearly, with $g(80) = 0.8$. The component strengths are assumed to degrade at the same rate following maintenance, i.e. the updating of probabilistic strength models is not considered for the simple illustration. The degradation function and failure probabilities with these strategies are illustrated in Figs.7.11 and 7.12, respectively. At the time of inspection and maintenance, the failure probability changes its slope; this change is more distinct when the component is inspected more thoroughly. If $F(100) \leq 0.028$, both strategies would be acceptable in terms of risk and the choice would be on the basis of cost as described in the following section.

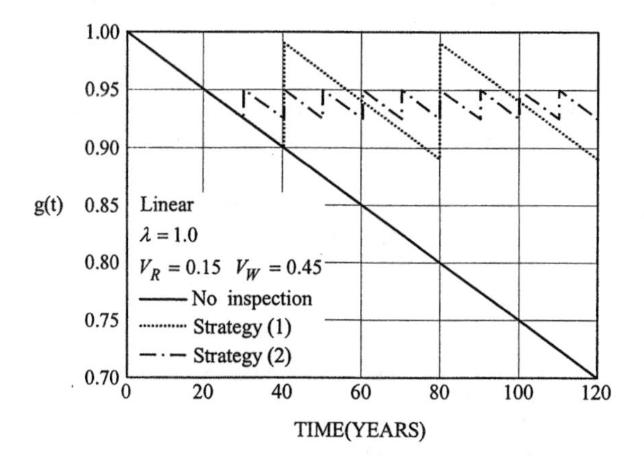

Figure 7.11: Degradation function with inspection/maintenance

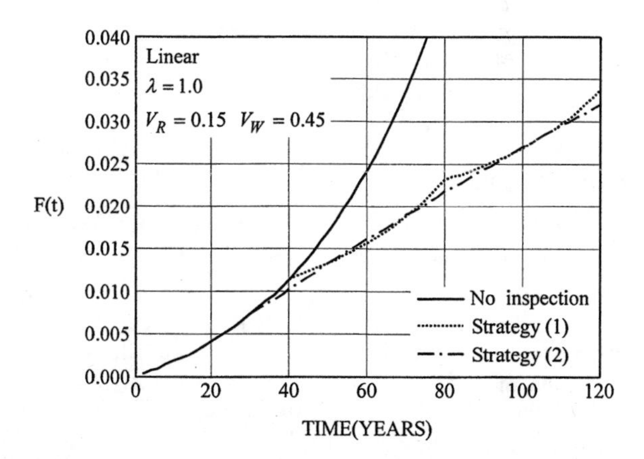

Figure 7.12: Failure Probability with inspection/maintenance

7.4.2 OPTIMUM IN-SERVICE INSPECTION AND REPAIR STRATEGIES

Costs of inspection and repair may be a significant part of the overall lifetime cost of a structure. These costs should be balanced by the benefits to be gained, both in economic and reliability terms. There are tradeoffs between the extent and accuracy of inspection, required level of reliability, and cost. One can perform this tradeoff by defining an objective function that takes into account the construction cost, cost of failure, cost of inspection, and costs of repair. The failure probability of a component during its service life, $F(t)$, provides one of several constraints on the optimization.

To design an optimum inspection/maintenance program for an existing structure, the following optimization problem must be solved:

$$\text{Minimize } C_T = C_{ins} + C_{rep} + C_f \cdot F(t_L)$$
$$\text{Subject to} F(t_L) < P_{f_{Target}} \tag{7.21}$$

in which C_{ins} = inspection cost, C_{rep} = repair cost, C_f = loss due to failure, and $P_{f_{Target}}$ = established target failure probability during service life, t_L. Other possible constraints might include the minimum and maximum time intervals between inspections, and the minimum threshold value of detection of the NDE method that is available.

Some studies have been done to determine optimal inspection/maintenance strategies for steel structures subjected to fatigue, assuming that a component is replaced if the intensity of detected damage exceeds a critical value (e.g., Thoft-Christensen and Sorensen, 1987). The cost of repair (i.e., cost of replacement) was assumed to be constant in these studies. However, a component may not be replaced; instead, only detected damages might be repaired. In this case, the cost of repair would be a function of damage intensities to be repaired. This aspect also should be considered in designing an optimum strategy.

The variables in the optimization are the inspection times, the extent of inspection and the anchor point of the NDE device, $\mu_{X_{th}}$, and the effectiveness or quality of repair. C_{ins} depends on the quality and extent of inspection. C_{rep} depends on the area and extent of damage to be repaired. C_f is in general difficult to assess in absolute terms. However, sensitivity studies can be conducted by varying the relative intensity of these costs. Such a study for an existing concrete structure without considering construction cost show that when failure costs dominate over other costs and the aging mechanism causes essentially linear degradation in strength over time, the optimal policy is to perform in-service inspection and repair at essentially uniform intervals (Mori and Ellingwood, 1994a,b).

When designing a new structure, the construction cost, C_I, is introduced as a part of total cost, C_T, in Eq.(7.21) and the probabilistic characteristics of the initial strength and even the degradation function would be the variables in the optimization and the problem would be considerably more complex.

CONCLUSIONS

Structural components may degrade with time. Such degradation would not be a problem for a structure with relatively short service life. However, when extending the service life of an existing structure and when designing a new structure with longer service life, it may impact on the safety of a structure over time and should be taken into account.

Reliability-based methodologies to evaluate time-dependent reliability of components and systems were described, and the sensitivity of the reliability to various parameters of strength degradation was illustrated using simple examples. The reliability is sensitive to the choice of initial component strength and strength degradation models, including induction time. Degradation characteristics and load process statistics must be identified for a particular facility to utilize the methodology in realistic assessment of durability.

The application of time-dependent reliability analysis to determine in-service inspection and repair strategies was illustrated. When there are limited resources for in-service inspection and repair, it is most effective to select a few safety-critical components and focus the monitoring effort on these components. The reliability analysis can be used to identify those components that are most significant from a risk viewpoint. Excluding the non-critical component from inspection and analysis, the cost of unnecessary monitoring can be minimized and the size of the system reliability analysis can be reduced.

When designing a new structure, one can choose a plan to design a component strong enough so that the deterioration would not cause any impact on the safety without in-service inspection and repair. One can also put resources to prevent deterioration so that in-service inspection and repair would not be required. To allocate a limited amount of resources along the service life of a structure rather than to spend them at once, it might be more reasonable to design a component to be less strong initially, but require periodic in-service inspection and maintenance. The decision should be made taking the possible change of the safety level and the economic status into account.

BIBLIOGRAPHY

AIJ. (1998) *Recommendations for limit state design of steel structures.* Architectural Institute of Japan. Tokyo, Japan. (in Japanese)

Albrecht, P. and A. H. Naeemi. (1984) "Performance of weathering steel in bridges." Report 272, Nat. Cooperative Highway Res. Program, July.

Ciampoli, M. "Reliability evaluation of existing structures: Updating technique to account for experimental data." *International Conference on Monitoring, Surveillance and Predictive Maintenance of Plants and Structures*, Taormina,

Italy, 1989, pp.149-158.

Clifton, J. R. and L. I. Knab. (1989). *Service life of concrete.* National Bureau of Standards, NUREG/CR-5466, U.S. Nuclear Regulatory Commission, Washington, D.C.

Enright, M. and D. Frangopol. (1998) "Service-life prediction of deteriorating concrete bridges." *J. Structural Engineering*, ASCE. Vol.124. No.3, Mar. pp.309-317.

Enright, M. and D. Frangopol. (1999) "Condition prediction of deteriorating concrete bridges using Bayesian updating." J. Structural Engineering, ASCE. Vol.125. No.10, Oct. pp.1118-1125.

Feddersen, C. E. (1970) "Evaluation and prediction of the residual strength of center cracked tension panels." ASTM STP No.486, pp.50-78.

Kameda, H. and T. Koike. (1975) "Reliability theory of deteriorating structures." *J. Structural Division*, ASCE. Vol.101, No.WT1, pp.295-309.

Kemp, M. E. (1987) "Atmospheric corrosion ratings of weathering steels – Calculations and significance." *Material Performance*, Vol.26, No.7, pp.42-44.

Melchers, R. E. (1987). *Structural reliability; analysis and prediction.* Ellis Horwood Ltd., West Sussex, England.

Melchers, R.E. (1990). "Search-based importance sampling." *Structural Safety*, Vol.9, pp.117-128.

Mori, Y. and B. Ellingwood. (1993a) "Time-dependent system reliability analysis by adaptive importance sampling." *Structural Safety*. Vol.12, pp.59-73.

Mori, Y. and B. Ellingwood. (1993b) "Reliability-based service life assessment of aging concrete structures." *J. Structural Engineering*, ASCE, Vol.119, No.5, pp.1600-1621.

Mori, Y. and B. Ellingwood. (1994a) "Maintaining reliability of concrete structures I: Role of inspection/repair." *J. Structural Engineering*, ASCE, Vol.120, No.3, pp.824-845.

Mori, Y. and B. Ellingwood. (1994b) "Maintaining reliability of concrete structures II: Optimum inspection/repair strategies." *J. Structural Engineering*, ASCE, Vol.120, No.3, pp.846-862.

Mori, Y and B. Ellingwood. (1995). "Reliability-based condition assessment of structures degrading due to environment and repeated loading." *Proc. of the 7th International Conference on Applications of Statistics and Probability in Civil Engineering*, Paris, France. pp.971-976.

Oehlers, D. J., A. Ghosh, and M. Wahab. (1995) "Residual strength approach to fatigue design and analysis." *J. Structural Engineering*, ASCE. Vol.121, No.9, pp.1271-1279.

Oswald, G. F. and G. I. Schueller. (1984). "Reliability of deteriorating structures," *Engineering Fracture Mechanics*, Vol.20, No.3, pp.479-488.

Somerville, G. (1986) "The design life of concrete structures." *The Structural Engineer*, Vol.64A, No.2, pp.60-71; Discussion, Vol.64A, No.9, 1986, pp.233-241.

Thoft-Christensen, P. and Sorensen, J.D. (1982). "Reliability of structural systems with correlation elements." *Applied Mathematical Modeling*, Vol.6, pp.171-178.

Thoft-Christensen, P. and Sorensen, J.D. (1987). "Optimal strategy for inspection and repair of structural systems," *Civil Engineering System*, Vol.4, pp.94-100.

Tuutti, K. (1982). *Corrosion of steel in concrete.* Swedish Cement and Concrete Research Institute, Stockholm, CBI-research 4:82.

Vesely, W. E. (1987) "Risk evaluations of aging phenomena: the linear aging reliability model and its extensions." *Proc. 14th Water Reactor Safety Information Meeting*, NUREG/CP-0082, Vol.3, February, pp.335-350.

Vesikari, E. (1988). *Service life of concrete structures with regard to corrosion of reinforcement.* Research Reports, Espoo, Finland.

Washa, G. W. and K. F. Wendt. (1975) "Fifty-year properties of concrete." *J. American Concrete Institute*, Vol.72, No.1, pp.29-38.

Washa, G. W., J. C. Saemann, and S. M. Cramer. (1989) "Fifty-year properties of concrete made in 1937." *ACI Materials Journal*, Vol.86, No.4, pp.367-371.

Zheng, R. and B. Ellingwood. (1999) "Stochastic fatigue crack growth in steel structures subjected to random loading." *Structural Safety*, Vol.20, No.4, pp.303-323.

Chapter 8

Soils and Foundations

By:
Jack Pappin, Director, Ove Arup & Partners Hong Kong Ltd

8.1 Introduction

This chapter discusses what can happen to foundations that may affect the safety of a tall building in its lifetime. It does not address how to design the foundations, as this information is available in many references (e.g. Tomlinson, 1995). Foundations in the context of this chapter include the items that support the structure, for example concrete pads, strip foundations and rafts, all types of piling or caissons, as well as basements beneath the superstructure constructed within the soil.

Items that are addressed in this chapter include the following:

Soil changes – this includes consolidation and creep and ground water variations that may occur for many years after the construction is complete

Wind loading – what factors need to be considered when assessing the effects of future wind loading.

Seismic loading – how can the soil type affect the dynamic response of the structure and apply load to the foundation system. The effects of liquefaction and slope instability are also addressed.

Adjacent construction – clearly a poorly controlled deep excavation adjacent to a high rise structure could have a significant impact on the existing structure. Methods of control and assessment of the impact on the existing structure are discussed.

8.2 Soil changes

Soil changes that can affect a completed structure generally involve movement of the ground. This can be caused by changes in the ground water regime that in turn leads to changes of the effective stress within the soil but for certain soil types can be the result of ongoing consolidation or creep.

8.2.1 Changes in the ground water regime

Changes in the ground water regime can result from many causes including changes in local ground water extraction, rupture of services and extreme climate conditions. In many cases the water table is a few metres below ground level and therefore the upper part of the basement structure requires virtually no waterproofing under normal conditions. Extreme weather or a ruptured water supply pipe can lead to significantly enhanced ground water levels and porewater pressures for a short period. Clearly this can lead to failure of the waterproofing (if it is not designed for this condition) or contribute to the failure of slopes that may exist in the immediate vicinity.

More extreme results can occur. In one case the basement retaining system in the upper few metres of a basement in a gravel material relied on contiguous bored piles to retain the soil. These piles had up to a 100mm gap between them and were quite adequate under normal conditions (below the water table an insitu concrete infill had been installed). A burst water main subsequently caused the water table to rise by several metres and this caused a gravel water mixture to flow between the piles into the basement. The buoyancy forces on the structure can also change significantly. This may lead to the basement floor slab to become over-stressed as it is spanning between columns and in extreme cases could lead to floatation of the entire structure especially for podium structures which often surround high rise buildings.

Many cities in the world have depressed ground water regimes due to water extraction. The central part of London in the United Kingdom, for example, has experienced ground water drawdown in the permeable soils beneath the London Clay for over a century. Recently however the quantity of pumping has significantly reduced and the pore water pressure in the underlying soils is increasing. Eventually this may give concern to the water proofing of basements which, at present, have no excess water pressure applied to them and therefore require minimal waterproofing measures (Simpson et al, 1989).

8.2.2 Changes in the effective stress in the soil

Changes in effective stress can arise from changes in the pore water pressure regime or from external loading from adjacent structures. This latter item is discussed in Section 5 of this chapter.

The previous section discusses several reasons for change in the ground water pressure in the soil where the pore water pressure is increased. This will cause a reduction in the effective stresses in the soil causing them to expand slightly. This expansion is generally small and generally the structure can cope with the resulting movement. A more alarming condition is where the structure is founded on piles that rely on soil friction for their support. The increase in pore water pressure will reduce the insitu horizontal effective stress in the soil thereby reducing their capacity. Generally there are large safety margins, which will cater for this extreme condition, but the designer should ensure the design is tolerant of this situation.

Problems can also arise as a result of reduction in the ground water pressure. The pore water pressure generally reduces because of additional water extraction in the vicinity often as a result of nearby construction activities. This reduction in porewater pressure causes an increase in the effective stress in the soil leading to settlement of the soil. In Hong Kong, for example, there have been many instances where ground water drawdown has caused problems for services etc. and the authorities' demand that detailed assessments of these affects be made before construction activities commence. Where deep extraction is occurring this leads to settlement of all soil layers above that affected by the drawdown. When this condition occurs where deep piles or caissons are present very large negative skin friction forces can be applied to the pile. This will lead to settlement of the structure the magnitude of which will depend on the founding condition of the pile. An important consideration is the compression capacity of the pile itself. If sufficient negative skin friction is generated the pile it may suffer a compression failure.

8.2.3 Consolidation and creep of the soil

Fine-grained soils such as clays and fine silts can take many years to respond to a change in applied pressure. This effect is referred to as consolidation and is of two types. The first is primary consolidation that results from the pore water being slowly squeezed out from between the soil particles and in turn leads to increased effective stresses and settlement as described above. The second is secondary consolidation, which is a creep phenomenon that can continue for many years after the primary consolidation, is complete.

Large consolidation effects are common in areas of recent reclamation over soft fine-grained soils and must be allowed for in design. From a design perspective this reduces to ensuring all service connections and joints between piled structures and adjacent hard-standings are designed to cope with a large differential settlement between the structure and the surrounding soil. In many cases a void will develop beneath the lowest floor slab, which is designed to be suspended between the column supports. Adequate provisions are required to allow long-term ventilation etc. Consolidation of the soil will also often lead to negative skin friction forces being applied to piles and caissons.

Even in stiff clay soils long-term consolidation settlement can occur. There are many examples in London for example where settlement of tall buildings has continued for a few years (Butler 1975). A special form is soil heave that arises in fine-grained soils after the stresses in the soil are reduced. In London Clay, for example, there are many instances where the soil has heaved for many years after completion of the structure. The Shell Centre in London is an example where the new structure weighed considerably less than the original soil that was excavated to form the basement. This structure continued to rise out of the ground for several decades and rose by a total of over 60mm (Burford 1988). Another situation is the long-term soil pressure acting up on the underside of the deepest basement floor slab. If the slab is stiff and cast directly onto the soil then eventually the earth pressure will increase to be near that existing in the soil at that depth prior to the excavation. Voids are often left in place beneath the lowest slab to eliminate this problem (e.g. British library, Loxham et al 1989).

Soil creep is the condition where the soil moves over time even though there is no change in the effective stresses in the soil. Usually it is the form of settlement of fine-grained soils as described above but it is also a well-known phenomenon in rockfill and other loosely placed granular fills (Pickles and Tosen 1998, Oldecop and Alonso 2001). Another area where soil creep is common is adjacent to the crests of man made slopes cut from the natural terrain. In these circumstances the soil movements have a large horizontal component that spreads towards the cut slope. This can lead to lateral forces being applied to many piles supporting the structure. In some circumstances piles are sleeved to enable the soil move past them by a small amount without applying load to the piles. It is difficult to maintain these systems however and generally it is preferable to design the piles to resist the imposed loads. The design method for this condition is similar to that used for earthquake induced lateral loading arising from soil movement and is discussed later.

8.3 Wind loading

One important concern with the long-term performance of a high rise structure subjected to extreme wind loading is to determine how the laterals load and the bending moments are transferred to the underlying foundation.

8.3.1 Without basement

For a building without a basement the load transfer is relatively simple, as the loads must directly transfer through the foundations. For a building founded on a raft, bearing capacity considerations must be made which explicitly allow for both the imposition of the lateral load as well as the overturning moment (see Figure 1). Vesic (1975) gives methods for doing this and these show the dramatic reduction in bearing capacity that can be found when significant lateral loading is present.

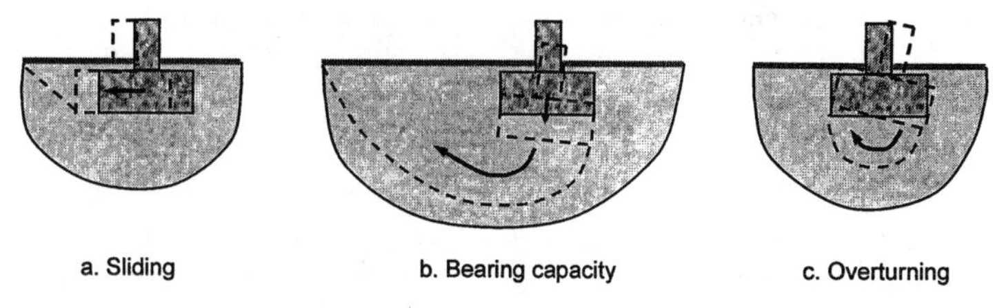

Figure 1 Soil failure modes for a pad or raft foundation

For other types of foundations including pads, strips and piles a soil structure interaction analysis will be required to quantify the lateral loading being applied to each of the foundation elements and the breakdown of the applied overturning moment. The moment
will be resisted by a combination of push pull forces from the outer foundation elements and restoring moments principally from core and shear wall foundations. Generally limit state codes of practice give good advice about the wind load combinations that need to be designed against. In particular, these codes have a load combination of minimum dead load combined with an ultimate wind loadcase (usually 1.4 to 1.6 times the nominal design wind load) to give an extreme uplift requirement for the foundations. Use of a 1.4 factor on a 50-year return wind load reduces the annual risk of wind load in extra-tropical wind regions to about 1 in 1,000, compared to 2% per annum for the 50-year return wind load itself. Working stress design codes are potentially quite unsafe in these conditions because if the nominal design wind load does not produce uplift, but is close to doing so, there may be no safety reserve for more extreme but credible design scenarios. This applies equally to the superstructure design.

When assessing the suitability of the foundation system it is important that it is proportioned such that excessive plastic or permanent deformation sufficient to cause misalignment of the whole structure is avoided. Depending on the soil, it may be important that the duration of an extreme wind storm is likely to be in the order of 12 hours and that extreme load will only occur for a few seconds within that period. The effects of cyclic loading during storms of less than maximum amplitude may also need to be considered as shake down may occur after many cycles of loading.

The analysis of piles subjected to lateral loading, can generally be carried out using a beam on spring approach and various computer codes exist to do this. For large pile groups allowance must be made for group effects, which may significantly reduce the apparent soil lateral stiffness.

For tall structures, the flexibility of foundations may be significant compared to the structure and this may need to be included in the dynamic modelling used to calculate the wind loads. It is important that appropriate short-term low strain dynamic stiffness is used in these assessments.

8.3.2 With basement

For a building with a basement the load transfer is generally quite complex as the wind loads transfer through the basement walls in addition to through the foundations. Again a soil structure interaction analysis will be required to quantify the lateral loading and bending moment being applied to each of the foundation elements and the walls. The basement floor slabs as well as the ground floor slab may be required to transfer a substantial part of the wind loading to the perimeter basement walls.

8.4 Seismic loading

Superficially the loading applied to the structure from seismic loading has many parallels to wind loading. However there are fundamental differences between these two types of load in that the seismic loading on the superstructure is applied by way of the foundation soil system.

Many codes of practice, including Eurocode 8 and UBC 1997, specify an ultimate load that originates from a ground motion having a 10% chance of being exceeded in a 50 year period (equivalent to a motion having a return period of 475 years or an annual risk of about 1 in 500). Performance Based Design methods such as FEMA 356 (2000), a prestandard and commentary for the seismic rehabilitation of buildings, specifies several levels of ground motion. The buildings should be checked to maintain life safety when subjected to a ground motion having a return period of 475 years and not to collapse when subjected to the maximum credible event, defined as having a 2% chance of being exceeded in 50 years or a return period of 2,475 years. Additional requirements, which may be desired by the owners, are for essentially no damage under frequent ground motions having a return period of 72 years and repairable damage with a ground motion having a return period of 225 years (50% and 20% chance of being exceeded in 50 years respectively). The IBC 2000 goes someway towards this goal by specifying that the design ground motion used to derive the ultimate load should be based on 2/3rds of the 2,475 year return period motion.

8.4.1 Effect of soil type on ground motion

Virtually all well established codes of practice contain direct allowance for the soil types present under the site. This is to allow for the now well known effects of site response which is effectively the soil deposit itself responding or resonating to the underlying dynamic earthquake induced rock motion. The earthquake that affected

Mexico City on September 19th 1985 was an extreme example of this amplification where the upper 30m thick, almost elastic, soft clay layer resonated with the input motion. The resonance period of the clay layer was about 2 seconds, which combined with the considerable long period energy present in the incoming earthquake ground motion, caused an increase of about 20 times in terms of force applied to a structure having a natural period of about 2 seconds. Most of the 10 to 15 storey buildings suffered significant damage as a result.

For a level site the effect of site response can be evaluated quite easily using one-dimensional codes such as SHAKE (Schnabel et al, 1972) or *Oasys* SIREN (Pappin, 1991). These codes model the soil allowing for the well known phenomenon that the soil stiffness reduces as the strain level increases which in turn causes the internal damping within the soil to increase significantly with increasing strain level. In an attempt to produce a simpler method for everyday design use, various systematic studies have been carried out to determine the effects of different soil types on the ground level earthquake ground motion (e.g. Heidebrecht and Rutenberg 1993, Borcherdt 1994). The most advanced method currently available is that incorporated into the recent IBC 2000 code of practice for the United States of America. This code classifies the soil profile in terms of its average shear wave velocity in the upper 30m (see Table 1). The ground motion amplification factors recommended by the code are shown in Figure 2. Two sets of factors are given, one set for short period ground motion and one set for long period ground motion. The factors in both sets are dependent on the level of input ground motion which is defined in terms of the bedrock response spectral values at the 0.2 or 1.0 second for short and long period motion respectively. Figure 2 shows that the short period amplification factors are more dependent on the level of motion than the long period factors. It is seen that loose/soft soil sites can experience significant amplifications at small earthquake ground motion. The amplification for Soil Profile Type E at low acceleration values gives rise to the same level of long period ground motion for a hard rock site having a basic acceleration coefficient that is four times greater.

Soil Profile Type	Shear wave velocity (m/sec)	SPT N value	Undrained shear strength (kPa)
A - Hard rock	>1,500	-	-
B - Weak to medium rock	750 - 1,500	-	-
C - Dense stiff soil	375 - 750	> 50	> 100
D - Medium dense firm soil	180 - 375	15 - 50	50 - 100
E - Loose soft soil	< 180	< 15	< 50
F - Deep soft soils that require site specific investigations			

The upper 30m of the soil profile are considered

Table 1 UBC soil classification system

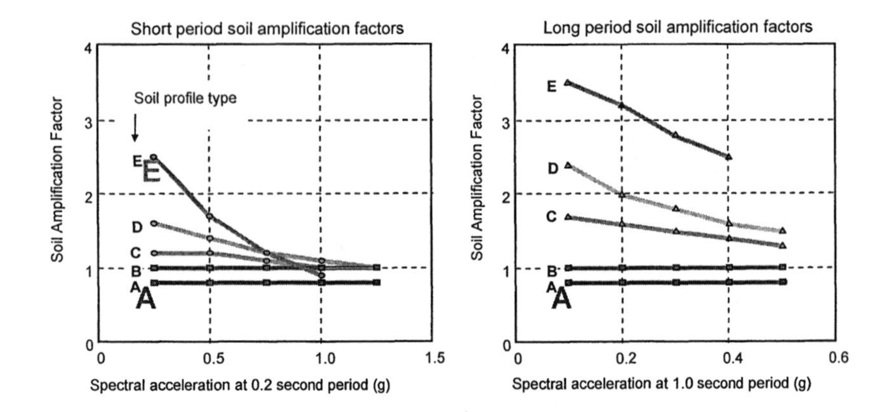

Figure 2 IBC 2000 Soil amplification factors for different levels of input motion

The depth of soil to bedrock is also important in the consideration of site response effects. For instance relatively thin medium dense sand deposits can exhibit significant amplification (Arup, 1993). This effect is not generally covered by codes however.

Site response effects become more complex in the situation of deep soil filled valleys where a one-dimensional analysis is not really sufficient and a two or even three-dimensional analysis is required to adequately capture the site effects. Rassem et al (1997) gives good guidance on how to estimate these effects to see if a more detailed analysis is likely to be necessary. Another example of two-dimensional effects is ridge effects where it is observed those ridges or other locations where there is a significant change of slope give rise to s significant amplification of the ground motion. The French Code of Practice (AFPS, 1990) explicitly allows for these effects.

Soil conditions that vary across the site are an additional effect because it is likely that the founding system will have a higher lateral stiffness under one side of the structure than under the other side. This will lead to the building experiencing a torsional response because the centre of mass of the structure will not coincide with the centre of stiffness of the foundation. It may be necessary to install a shear wall in the ground, usually in the form of a diaphragm wall, in the weaker area to stiffen this part of the foundation.

8.4.2 Effects of the foundation soil system on structural response

As described above, the earthquake force that the structure needs to resist is a function of the natural period of the structure. For structures founded on soil the effect of the underlying foundation compliance will lengthen this fundamental period. In addition the existence of the structure will affect the ground motion it experiences at its base. Both these effects are referred to as dynamic soil-structure interaction (DSSI).

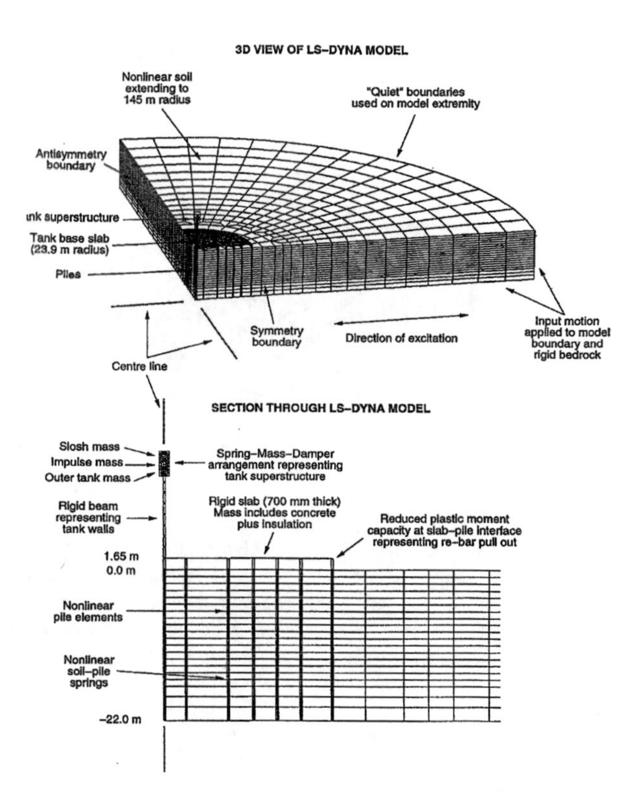

Figure 3 Finite element mesh for DSSI in LS-DYNA (Lubkowski et al, 2000)

DSSI can have a significant effect on the dynamic response of a structure. However rigorous techniques for incorporating DSSI require elaborate computations. The most common method in use is to use a three-dimensional analysis tool modelling the soil and superstructure simultaneously. Figure 3 shows a finite element mesh used in the explicit non-linear time stepping program LS-DYNA for example. Simplified methods exist for raft footings etc. and Wolf (1994) and Roesset (1980) describe these. Generally the simplified methods use frequency dependent compliance functions (e.g. Veletsos and Wei, 1971) that limit the analysis to linear frequency domain analysis methods (see Booth et al, 1988).

8.4.3 Assessment of foundation adequacy

Having established the earthquake force that the structure needs to resist, the foundation design considerations are very similar to those for wind load. The load paths for the lateral and overturning moments need to be established and the foundation system checked to ensure that the foundation system is proportioned such that excessive plastic or permanent deformation sufficient to cause misalignment of the whole structure is avoided. An extreme and unusual example of a rotational failure of a raft foundation in Mexico City in the 1985 event is shown in Plate 1.

Plate 1 Rotational failure of a raft foundation in the 1985 Mexico City earthquake (EEFIT 1986)

8.4.4 Effects of seismic ground motion on piles

Where earthquake and wind loading differ fundamentally is their effect on deep foundations. Wind loading applies loading only to the top of the foundations, whereas seismic ground motion can directly load deep foundations. Determining the effect of seismic motion on the piles involves considering the lateral soil load on the piles in addition to the lateral seismic load from the superstructure. Figure 4 illustrates the nature of the problem. The soil near the top of the pile is loading the pile in the same direction as the superstructure load. For piles that are flexible relative to the soil, the effects of soil movement will be minimal. For stiff large diameter piles or caissons however the loading from the soil may be significant and allowance must be made for this when checking the pile design. A rigorous procedure for this is to use a non-linear three- dimensional three-dimensional finite element analysis such as LS-DYNA. A robust simplified procedure for doing this described in Pappin et al (1998) and comprises setting up a model of the pile as a beam on elastic/plastic springs and applying a free field soil displacement to the ends of the springs as illustrated in Figure 5. The soil displacement can generally be obtained directly from a one dimensional site response analysis such as *Oasys* SIREN.

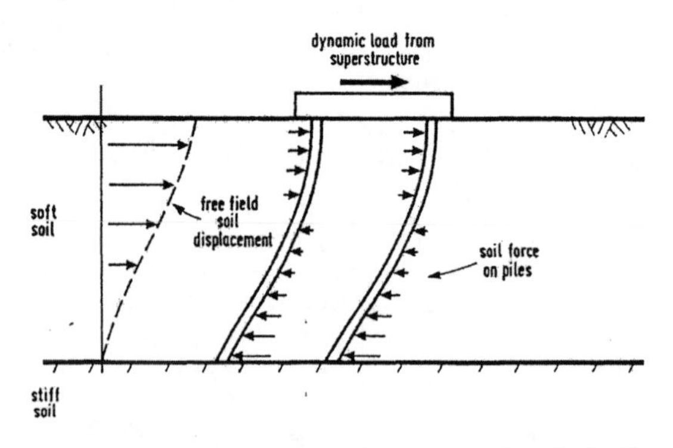

Figure 4 Lateral loads on piles subjected to earthquake loading

This type of analysis often shows that the piles experience significant bending moments where they pass through a soil boundary between a hard and soft soil and also at or near the head of the pile. A load combination rule using SRSS or a 100% - 30% rule (UBC, 1997) needs to be used to combine the soil and structural loading to avoid being over-conservative.

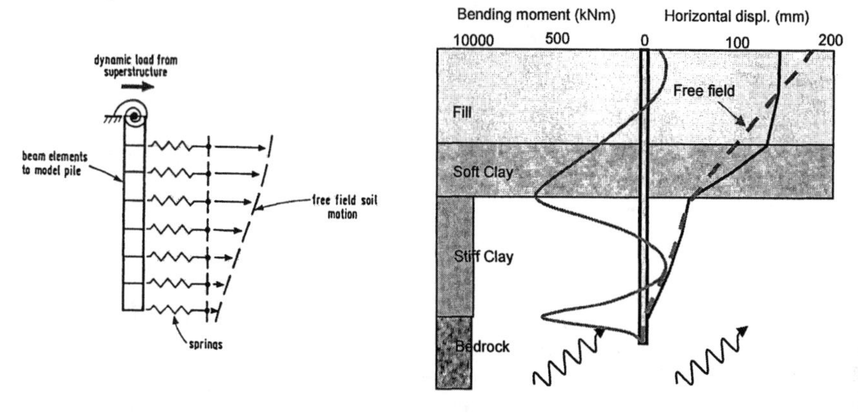

Figure 5 Diagrammatic model of pile soil system and typical results. The rotational spring models the restraint from the superstructure. The horizontal springs, which are elastic/plastic, model the resistance provided by the soil to the pile moving horizontally through the soil.

For good behaviour of the foundation piles in future design level seismic events it is necessary that the piles be properly detailed in the locations of high bending moments so that they do not experience a brittle type of failure. Often an elastic design procedure is used for piles such that in the design seismic ground motion they do not experience any distress. Even if this rather conservative decision is made it is important that good detailing rules are used to ensure the pile will have a good possibility of continuing to fulfil its load carrying function even after experiencing a major extreme seismic loading event.

Raking piles are a special case that must be discussed. Occasionally they are used to stiffen the lateral stiffness of the foundation system. This is an acceptable practice but because of the high lateral stiffness of the raking piles compared to the vertical piles they will carry almost the entire imposed lateral load. In addition they will cause the foundation system to be very stiff relative to the soil and the soil will therefore impose a much greater lateral load onto the foundation as a whole. This additional soil load will also be transferred to the raking piles as illustrated in Figure 6. Inevitably very large loads often get applied to the raking piles and there are several instances where they have failed under seismic loading. Some countries prohibit the use of raking piles as a result (e.g. Turkish Code 1998).

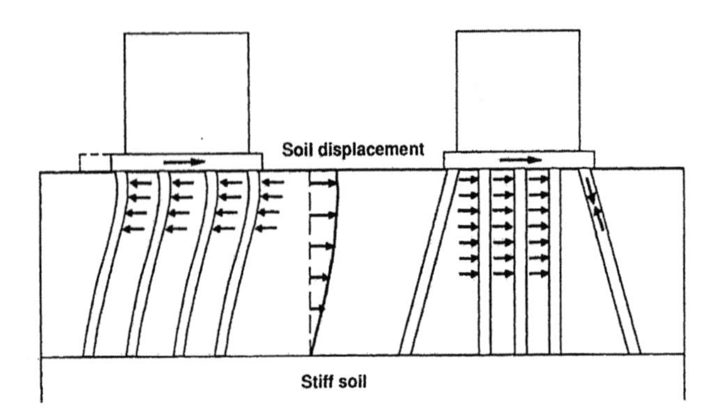

Figure 6 The effect of raking piles on pile group deformations

8.4.5 Effects of seismic ground motion on basements

There are various ways of assessing the seismic loading experienced by basement structures. The simplest in concept is to use a large three-dimensional finite element method such as that illustrated previously in Figure 3. For a long basement structure this can be reduced to a two-dimensional study preferably using a dynamic analysis but a good approximation can be achieved by imposing the lateral soil displacement from the one-dimensional site response analysis to a static two-dimensional finite element analysis. Some form of horizontal body force must also be applied to the soil mass, so that the soil profile, in the absence of the structure, will deform by the correct amount. This is the procedure recommended by the Japanese Ports and Harbours Authorities (JSCE, 1992) and essentially follows the same principal as that shown in Figure 5.

An additional consideration is the maximum earth pressure that can be applied to the external wall of the basement. A basement will generally appear to the soil to be essentially a rigid wall and the formula proposed by Wood in 1973 is commonly used for this situation. This states that the force increase per unit length of wall is equal to the peak ground acceleration times the density of the soil times the square of the wall depth and that the centre of action of this force is about 40% down the wall (see Kramer 1996). Wu and Finn (1999) show a refined method for this assessment.

8.4.6 Liquefaction

When subjected to cyclic loading or ground shaking loose granular soils tend to reduce in volume by way of the soil grains moving relatively to form a closer or denser state of packing. For dry soils this results in a denser material. For soils that are saturated with water, or nearly saturated, the pore water cannot escape from between the soil particles quickly enough leading to a rise in pore water pressure. In many instances the pore water pressure increases to equal the total vertical stress that means the effective stresses between the soil grains reduce to zero and the soil essentially behaves as a dense liquid, a state referred to as liquefaction.

Liquefaction is observed at many locations during most large earthquake events. It can take the form of the upper few metres of soil completely liquefying or a layer of soil liquefying beneath a crust of soil that does not liquefy. In this latter situation evidence of the underlying liquefaction is seen as 'sand boils' created by liquefied sand reaching the surface by way of small diameter pipes forcing their way through the overlying material. There are several established ways of determining the likelihood of liquefaction (Seed et al 2001).

The effects of liquefaction on buildings depend on the site-specific conditions and often raft and pad foundations suffer extreme bearing capacity failure leading to the buildings sinking by up to several metres and invariably leaning at large angles. It is interesting to note that the liquefied soil tends to insulate the building from the strong motion and it is commonly observed that buildings on liquefied soil suffer much less structural collapse than nearby structures on firmer soil. They often, however, are left standing at an extreme angle and are not reusable.

While buildings on piles usually fare much better than those on pads or rafts, the piles themselves may have to withstand large lateral displacements and also preserve their load carrying ability. It is apparent that liquefied soil still retains some small strength (Stark and Mesri 1992) and soil with this strength and corresponding stiffness can be modelled in a site response analysis to determine the likely relative displacement imposed on the piles by non liquefied soil above and below the liquefied layer. The procedure described previously in Figure 5 can then be used to assess the pile behaviour.

An alternative is to perform some of ground treatment to prevent liquefaction occurring even if a large seismic event occurs. These treatments include grouting, densification and improving the drainage within the liquefiable soil. A comprehensive discussion is given in NCEER (1997).

The above discussion is directly applicable to level sites. If the site is sloping or if it is near the edge of an embankment then it is highly likely that lateral spreading or a flow slide will occur leading to large horizontal movements of the soil. Kramer (1996) gives a good summary of how to evaluate such effects. Generally buildings on pads or rafts would be unable to survive this situation but it is possible to design piles that can resist such movements usually by allowing the soil to flow past them.

8.4.7 Slope instability

Slope failures are frequently observed in large earthquakes in regions of mountainous terrain. The assessment of the likelihood of such events is usually based on the method published by Newmark in 1965. This method calculates the static horizontal ground acceleration that is required to cause the slope to be on the verge of failing. This 'critical' acceleration is then compared to the peak ground acceleration that is expected at the site. If the 'critical' acceleration is not less than half of the applied peak acceleration then the ground movements are likely to be small, in the order of a few millimetres. For more onerous situations where the 'critical acceleration' is smaller then various authors have published methods to estimate the likely ground displacement (e.g. Sarma, 1975).

When assessing the safety of a building to withstand these movements the design must study the ability of the piles to tolerate the soil movements. This is illustrated in

Figure 7 and is similar in principal to the method of assessing site response effects on piles.

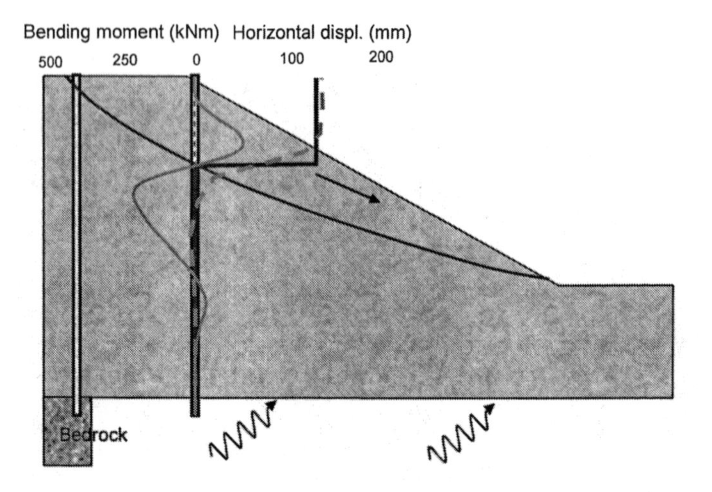

Figure 7 Schematic of a slope moving past a line of piles

If failure of the slope could lead to instability of the piles or foundation system an important consideration is to directly consider the implications of an extreme earthquake event. If the extreme event could lead to a slope failure then this must be considered directly. It is not adequate to design for a more frequent ground motion, such as the 475 year return period motion, and include detailing rules to cater for a more extreme situation, such as the 2,475 year return period motion. It is necessary to directly address the design for this extreme event.

8.4.8 Fault rupture

The earthquakes in Turkey and Taiwan in 1999 showed extensive damage arising from fault rupture. In both cases buildings straddling the ruptured part of the fault were
Extensively damaged whereas buildings only a few metres away survived unscathed. The lesson from these, and other examples, is that it is very important that the building does not straddle an active fault. While the locations of active faults are clear after a major earthquake specialised geological studies are generally required to determine the locations of such features.

8.5 Adjacent Construction

8.5.1 Ground movement

As discussed in Section 2 ground water drawdown is a common phenomenon arising from adjacent construction and it is necessary that the design of buildings allows for a reasonable variation in the ground water table after the structure is complete. Ground movements can arise from other activities related to adjacent construction however. For example if a deep basement is to be constructed using diaphragm walls the surrounding soil, including that under the building being considered, will experience movement due to the installation of the wall and also because of the subsequent excavation. Movements of the soil under the building may also arise from adjacent tunnelling work or from the construction of a structure that imposes a heavy load to the ground.

Many analysis programs are available to predict the size of these ground movements. They generally use a finite element approach and comprise some type of soil model. The most common soil model is an elastic perfectly plastic approach that characterises the soil by elastic constants until some failure stress condition is reached. In the past few years however it has become clear that non-linear soil models that allow for high stiffness at low strains are necessary to better predict the ground movements. Various models are now in use and a good summary of this work can be found in Atkinson, 2000.

8.5.2 Construction control

Good site control is vital to minimise ground movements during construction. This is especially important for below ground construction where, in many cases, the work is not able to be inspected externally and for temporary excavation support works.

A common case where below ground construction leads to ground movement is the installation of diaphragm wall panels and bored piles. Some of these movements are inevitable, for example see Davies and Henkel (1980) that studies the installation of diaphragm walls. Other ground movements are very operator dependent however. For example in large bored piles excavated by temporary casing and grab the amount of ground loss can be greatly affected by whether the grab is operating above or below the toe level of the casing. Maintaining the level of drilling mud above the surrounding ground water level will often have a crucial effect on the amount of ground movement induced.

Large ground movements can be induced by different piling systems. For example rotary percussive down-the-hole hammers that operate by air flush can cause significant loss of ground around the piling. This is particularly noticeable when this system is used in areas of loose soil with a high water table because the air return tends to act as an air-lift system to carry water and soil from around the pile shaft to the soil surface.

Temporary support systems to excavations are an important area requiring good site control. A study recently carried out for the Hong Kong Government (GEO, 2002) showed that the site control of temporary supports is poor, especially when sheet piles are used as temporary supporting systems. These systems tend to use many levels of temporary propping that impede excavation progress. There is therefore a tendency to remove them at times of critical activities and some large failures have resulted. Control of the penetration of the sheet piles themselves was also observed to be poor. These problems are not unique to Hong Kong and good site control practices are essential at all construction sites.

8.6 References

AFPS (1990) "Recommendations AFPS-90 for the redaction of rules relative to the structures and installations built in regions prone to earthquakes", Association Francaise du Genie Parasismique, Saint-Remy-les-Chevreuse, France.

Atkinson J.H. (2000) "Non-linear soil stiffness in routine design" Geotechnique Vol. 50, 40th Rankine Lecture, pp487-508, UK.

Arup (1993) "Earthquake hazard and risk in the UK", Report for Dept. of the Environment, UK.

Booth E.D., Pappin J.W. & Evans J.J.B. (1988) "Computer aided analysis methods for the design of earthquake resistant structures: a review" Proceedings of Inst. of Civil Engineers, UK, Part 1, Vol. 84, pp 671-691, August.

Borcherdt R. D. (1994) "Simplified site classes and empirical formulations for site response dependent code provisions", Proc. NCEER/SEAOC/BSSC Workshop on site response during earthquakes and seismic code provisions, Univ. of Southern California, Los Angeles, Edited by Martin G. M., USA.

Burford D. (1988) "Heave of tunnels beneath the Shell Centre, London", Geotechnique Vol. 38 No. 1 pp 1959-1986, UK.

Butler F. G. (1975), "General report and state of the art review", Session 3, Proc. Conf. On Settlement of Structures, Cambridge, Pentech Press, London, UK, pp 531-578.

Davies R.V. & Henkel D.J. (1980) "Geotechnical problems associated with the construction of Chater Station", Proc. of Conference on Mass Transportation in Asia, Paper J3, Mass Transit Railway Corporation, Hong Kong.

EEFIT (1986) "The Mexican Earthquake of 19th September 1985". A field report by EEFIT, Publ. by SECED Inst. of Civil Engineers, London, UK.

ENV, Eurocode 8 (1996) "Design provisions for earthquake resistance of structures - Part 5: Foundations, retaining structures and geotechnical aspects", European Prestandard ENV 1998-5.

FEMA 356 (2000) "Prestandard and commentary for the Seismic Rehabilitation of Buildings", Federal Emergency Management Agency, Washington, DC, USA.

GEO (2002) "QRA of collapses and excessive displacements of deep excavations", Report No. 124, Geotechnical Engineering Office, Civil Engineering Department, The Government of the Hong Kong Special Administrative Region.

Heidebrecht A.C. & Rutenberg A. (1993) "Seismic site-dependent response spectra for building codes: a proposal", Soil Dynamics and Earthquake Engineering, Vol.VI, pp691-704, UK.

IBC (2000) International Building Code, International Code Council, Cal, USA.

JSCE (1992) "Earthquake resistant design features of submerged tunnels in Japan" Earthquake resistant design for civil engineering structures in Japan, compiled by Earthquake Engineering Committee, The Japan Society of Civil Engineers.

Kramer S.L. (1996) "Geotechnical Earthquake Engineering", Prentice-Hall Englewood Cliffs, NJ, USA.

Loxham R., Simpson B. & Gatenby N. E. (1989) "Ground instrumentation at the British Library, Euston". Proc. ICE Conf. Instrumentation in Geotechnical Engineering, Nottingham, UK.

Lubkowski Z., Pappin J.W. & Willford M.R. (2000) "The influence of dynamic soil structure interaction analysis on the seismic design and performance of an ethylene tank." Proc. 12th World Conf. on Earthquake Engineering, Auckland, Jan., NZ.

NCEER (1997) "Proceedings of the NCEER workshop on evaluation of liquefaction resistance of soils", Edited by Youd T.L. & Idriss I.M., Technical Report No. NCEER-97-0022, Dec., USA.

Newmark (1965) "Effects of earthquakes on dams and embankments", Geotechnique, Vol 15, No 2, pp139-160, UK.

Oldecop L.A. & Alonso E.E. (2001) "A model for rockfill compressibility", Geotechnique, Vol. LI, No 2, pp127-139, UK.

Pappin J.W. (1991) "Design of foundations and soil structures for seismic loading" in "Cyclic loading of Soils". edited by O'Reilly M.P. & Brown S.F. pp 306-366, Blackie, UK.

Pappin J.W., Ramsey J., Booth E. D. & Lubkowski Z.A. (1998) "Seismic response of piles: some recent design studies." Proc. of The Institution of Civil Engineers, Geotechnical Engineering, Vol. 131, pp23-33, Jan., UK.

Pickles A.R. & Tosen R. (1998) "Settlement of reclaimed land for the new Hong Kong International Airport", Proc. of The Institution of Civil Engineers, Geotechnical Engineering, Vol. 131, pp191- 209, Oct., UK.

Rassem M., Ghobarah A. & Heidebrecht A.C. (1997) "Engineering Perspective for the seismic site response of alluvial valleys", Earthquake Engineering and Structural Dynamics, -Vol 26, pp477-493, UK.

Roesset J., (1980) "Stiffness and damping coefficients of foundations", ASCE Special Publication, Dynamic Response of Pile Foundations - Analytical Aspects, pp1-30, USA.

Sarma S.K. (1975) "Seismic stability of earth dams and embankments", Geotechnique, Vol 25, No 4, pp743-761, UK.

Stark T.D. & Mesri G. (1992) "Undrained shear strength of liquefied sands for stability analysis", Journal of Geo. Eng. ASCE 118(11), pp1727-1747, USA.

Schnabel P., Seed H.B. & Lysmer J. (1971) "Modifications of seismograph records for effects of local soil conditions", Report No EERC 71-8, University of California, Berkley, USA.

Seed H.B., Cetin K.O., Moss R.E.S., Kammerer A.M., Wu J., Pestana J.M. & Reiner M.F. (2001) "Recent advances in soil liquefaction engineering and seismic site response evaluation", Proc. 4th Int. Conf. on Recent Advances in Geo. Eng. and Soil Dyn. and Symp. in honour of Prof. W.D. Liam Finn, San Diego, Cal., March 26-31, USA.

Simpson B., Blower T., Craig R. N. & Wilkinson W. B. (1989) "The engineering implications of rising groundwater levels in the deep aquifer beneath London". CIRIA Special Publication 89, UK.

Tomlinson M.J. (1995) "Foundation design and construction", 6th Edition, Longman Scientific & Technical.

Turkish Seismic Code (1998) "Specification for structures to be built in disaster areas, Part III – earthquake disaster prevention", Ministry of Public Works and Settlement, Government of Republic of Turkey.

UBC (1997) "Uniform Building Code, Volume 2, Chapter 16, Division IV: Earthquake Design", International Conference of Building Officials, Whittier, California, USA.

Veletsos, A.S. & Wei, Y.T. (1971) "Lateral and rocking vibration of footings", ASCE Journal of Soil Mechanics and Foundation Division, Vol 97, No SM9, pp227-248.

Vesic A.S. (1975) "Bearing capacity of shallow foundations" Foundation Engineering Handbook, 1st Edn. Winterkorn H.F. & Fang H.Y. Chapter 3, Van Nostrand Reinhold Co. Inc. NY, USA.

Wolf J.P. (1994) "Foundation vibration analysis using simple physical models", Prentice-Hall Englewood Cliffs, NJ, USA.

Wu G. & Liam Finn W.D. (1999) "Seismic lateral pressures for design of rigid walls", Canadian Geotechnical Journal, Vol. 36, pp509-522.

Chapter 9

ASSESSMENT OF EXISTING STRUCTURES

By:
Milan Holicky, Klokner Institute, Czech Technical University

9.1 Introduction

Assessment of existing structures is becoming a more and more important and frequent engineering task. Continued use of existing structures is of a great significance due to environmental, economic and socio-political assets, growing larger every year. General principles of sustainable development regularly lead to the need for extension of the life of a structure, in majority of practical cases in conjunction with severe economic constraints. These aspects are relevant above all to tall buildings that always constitute a great social and economic value. In particular principles of sustainable development and need for extension of the life constitute significant constraints imposed on rehabilitation of tall buildings. That is why assessment of existing structures often requires application of sophisticated methods, as a rule beyond the scope of traditional design codes. Nevertheless, apart from few national codes, three International Standards (ISO 2394 [1], ISO/CD 13822 [2] and ISO 12491 [3]), related to assessment of existing structures have been recently developed. Additional information may be found in a number of scientific papers and publications, for example in [4] and [5].

The approach to assessment of an existing structure is in many aspects different from that taken in designing the structure of a newly proposed building. The effects of the construction process and subsequent life of the structure, during which it may have undergone alteration, deterioration, misuse, and other changes to its as-built (as-designed) state, must be taken into account. However, even though the existing building may be investigated several times, some uncertainty in behaviour of the basic variables shall always remain. Therefore, similarly as in design of new structures, actual variation in the basic variables describing actions, material properties, geometric data and model uncertainties are taken into account by partial factors or other code provisions.

In general, an existing structure may be subjected to the assessment of its actual reliability in case of:

- Rehabilitation of an existing constructed facility during which new structural members are added to the existing load-carrying system;

- Adequacy checking in order to establish whether the existing structure can resist loads associated with the anticipated change in use of the facility, operational changes or extension of its design working life;
- Repair of an existing structure, which has deteriorated due to time dependent environmental effects or which has suffered damage from accidental actions, for example, earthquake;
- Doubts concerning actual reliability of the structure.

In some circumstances assessments may also be required by authorities, insurance companies or owners or may be demanded by a maintenance plan.

9.2 Principles and general framework of assessment

Two main principles are usually accepted when assessing existing structures:

- Currently valid codes for verification of structural reliability may be applied; historic codes valid in the period when the structure was designed should be used only as guidance documents.
- Actual characteristics of structural materials, actions, geometric data and structural behaviour should be considered: the original design documentation including drawings should be used as guidance documents only.

The first principle should be applied in order to achieve similar reliability level as in case of newly designed structures. The second principle should avoid negligence of any structural condition that may affect actual reliability (in favourable or unfavourable way) of a given structure.

Most of the current codes are developed assuming the concept of limit states in conjunction with the partial factor method. In accordance with this method, which is mostly considered here, basic variables are specified by characteristic or representative values. The design values of the basic variables are determined on the basis of the characteristic (representative) values and appropriate partial factors.

It follows from the second principle that a visual inspection of the assessed structure should be made whenever possible. Practical experience shows that inspection of the site is also useful to obtain a good feel for actual situation and state of the structure.

As a rule the assessment need not to be performed for those parts of the existing structure that will not be affected by structural changes, rehabilitation, repair, change in use or which are not obviously damaged or are not suspected of having insufficient reliability.

In general, the assessment procedure consists of the following steps (see the flow chart in Annex A to this Chapter):

- Specification of the assessment objectives required by the client or authority;

- Scenarios related to structural conditions and actions;
- Preliminary assessment:
  - study of available documentation;
  - preliminary inspection;
  - preliminary checks;
  - decision on immediate actions;
  - recommendation for detailed assessment;
- Detailed assessment:
  - detailed documentary search;
  - detailed inspection;
  - material testing and determination of actions;
  - determination of structural properties;
  - structural analysis;
  - verification of structural reliability;
- report including proposal for construction intervention;
- repeat the sequence if necessary.

When the preliminary assessment indicates that the structure is reliable for its intended use over the remaining life a detailed assessment may not be required. Conversely if the structure seems to be in dangerous or uncertain condition immediate interventions and detailed assessment may be necessary.

9.3 Investigation

Investigation of an existing structure is intended to verify and update the knowledge about the present condition (state) of a structure with respect to a number of aspects. Often, the first impression of the structural condition will be based on visual qualitative investigation. The description of possible damage of the structure may be presented in verbal terms like: 'unknown, none, minor, moderate, severe, destructive'. Very often the decision based on such an observation will be made by experts in purely intuitive way.

A better judgement of the structural condition can be made on the basis of (subsequent) quantitative inspections. Typically, assessment of existing structures is a cyclic process when the first inspection is supplemented by subsequent investigations. The purpose of the subsequent investigations is to obtain a better feel for the actual structural condition (particularly in the case of damage) and to verify information required for determination of the characteristic and representative values of all basic variables. For all inspection techniques, information on the probability of detecting damages if present, and the accuracy of the results should be given.

The statement from the investigation contains, as a rule, the following data describing

- Actual state of the structure;
- Types of structural materials and soils;
- Observed damages;
- Actions including environmental effects;

- Available design documentation.

A proof loading is a special type of investigation. Based on such tests one may draw conclusions with respect to:

- The bearing capacity of the tested member under the test load condition;
- Other members;
- Other load conditions;
- The behaviour of the system.

The inference in the first case is relatively easy; the probability density function of the load bearing capacity is simply cut off at the value of the proof load. The inference from the other conclusions is more complex. Note that the number of proof load tests needs not to be restricted to one. Proof testing may concern one element under various loading conditions and/or a sample of structural elements. In order to avoid an unnecessary damage to the structure due to the proof load, it is recommended to increase the load gradually and to measure the deformations. Measurements may also give a better insight into the behaviour of the system. In general proof loads can address long-term or time-dependent effects. These effects should be compensated by calculation.

9.4 Basic variables

In accordance with the above-mentioned general principles, characteristic and representative values of all basic variables shall be determined taking into account the actual situation and state of the structure. Available design documentation is used as a guidance material only. Actual state of the structure should be verified by its inspection to an adequate extent. If appropriate, destructive or non-destructive inspections should be performed and evaluated using statistical methods.

For verification of the structural reliability using partial factor method, the characteristic and representative values of basic variables shall be considered as follows:

(a) Dimensions of the structural elements shall be determined on the basis of adequate measurements. However, when the original design documentation is available and no changes in dimensions have taken place, the nominal dimensions given in the documentation may be used in the analysis.

(b) Load characteristics shall be introduced with the values corresponding with the actual situation verified by destructive or non-destructive inspections. When some loads have been reduced or removed completely, the representative values can be reduced or appropriate partial factors can be adjusted. When overloading has been observed in the past it may be appropriate to increase adequately representative values.

(c) Material properties shall be considered according to the actual state of the structure verified by destructive or non-destructive inspections.

When the original design documentation is available and no serious deterioration, design errors or construction errors are suspected, the characteristic values given in original design may be used.

(d) Model uncertainties shall be considered in the same way as in design stage unless previous structural behaviour (especially damage) indicates otherwise. In some cases model factors, coefficients and other design assumptions may be established from measurements on the existing structure (e.g. wind pressure coefficient, effective width values, etc.).

Thus reliability verification of an existing structure should be backed up by inspection of the structure including collection of appropriate data. Evaluation of prior information and its updating using newly obtained measurements is one of the most important steps of the assessment.

9.5 Evaluation of inspection results

Using results of an investigation (qualitative inspection, calculations, quantitative inspection, proof loading) the properties and reliability estimates of the structure may be updated. Two different procedures can be distinguished:

(1) Updating of the structural failure probability.
(2) Updating of the probability distributions of basic variables.

Direct updating of the structural reliability (procedure (1)) can be formally be carried out using the following basic formula of probability theory:

$$\mathrm{P}(F|I) = \frac{\mathrm{P}\left(F \cap I\right)}{\mathrm{P}(I)} \qquad (1)$$

where P denotes probability, F local or global failure, I inspection information, and $\cap$ intersection of two events. The inspection information I may consist of the observation that the crack width at the beam B is smaller than at the beam A. An example of probability updating using equation (1) is presented in Annex B to this Chapter.

The updating procedure of a univariate or multivariate probability distribution (procedure (2)) is given formally as:

$$\mathrm{f}_X(x|I) = C\,\mathrm{P}(I|x)\,\mathrm{f}_X(x) \qquad (2)$$

where $\mathrm{f}_X(x|I)$ denotes the updated probability density function of X, $\mathrm{f}_X(x)$ denotes the probability density function of X before updating, X a basic variable or statistical parameter, I inspection information, C normalising constant, and $\mathrm{P}(I|x)$ likelihood function.

An illustration of equation (2) is presented in Figure 1. In this example updating leads to a more favourable distribution with a greater design value x_d than the prior design value x_d. In general, however, the updated distribution might be also less favourable than the prior distribution.

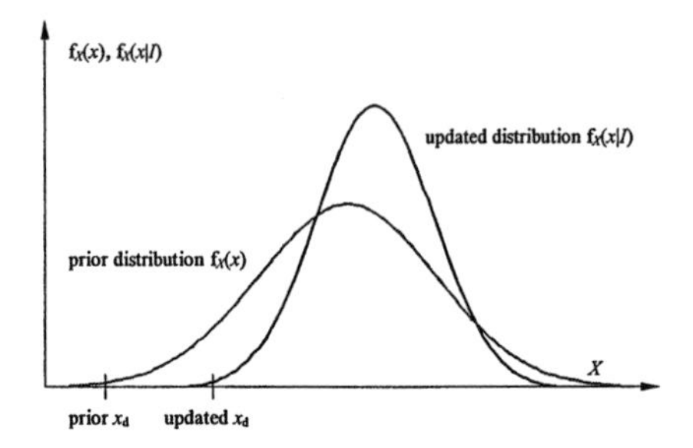

Figure 1. Updating of probability density function for an expected variable *X*.

The updating procedure can be used to derive updated characteristic and representative values (fractiles of appropriate distributions) of basic variables to be used in the partial factor method or to compare directly action effects with limit values (cracks, displacements). The Bayesian method for fractile updating is described in Annex C to this Chapter. More information on updating may be found in ISO 12491 [3].

Once the updated distributions for the basic variables $f_X(x)$ have been found, the updated failure probability $P(F|I)$ may be determined by performing a probabilistic analysis using common method of structural reliability for new structures. Symbolically it can be written

$$P(F|I) = \int_{g(x)<0} f_X(x|I)\,dx \tag{3}$$

where $f_X(x|I)$ denotes the updated probability density function and $g(x) < 0$ denotes the failure domain ($g(x)$ being the limit state function). It should be proved that the probability $P(F|I)$, given the design values for its basic variables, does not exceed a specified target value.

A more practical procedure is to determine updated design values for each basic variable (procedure (2)). For a resistance parameter *X*, the design value can be obtained using operational formula of ISO 2394 [1]. For normal and lognormal random variable it holds

$$x_d = \mu(1-\alpha\beta V) \tag{4}$$

$$x_d = \mu\exp(-\alpha\beta\sigma - 0.5\sigma^2) \tag{5}$$

where x_d is the updated design value for X, μ updated mean value, α probabilistic influence coefficient, β target reliability index, V updated coefficient of variation, and $\sigma^2 = \ln(1+V^2)$.

The value of the target reliability index β is discussed in ISO/CD 13822 [2], the values of α can be taken equal to those commonly used for new structures (0.7 for the dominating load parameter, 0.8 for the dominating resistance parameter and 0.3 for non-dominating variables according to ISO 2394 [1]).

As an alternative to procedure (2), one might also determine the characteristic value x_k first and calculate the design value x_d by applying the appropriate partial factor γ_m:

$$x_d = x_k / \gamma_m \tag{6}$$

For normal and lognormal random variable X the characteristic value x_k then follows as

$$x_k = \mu(1 - kV) \tag{7}$$
$$x_k = \mu \exp(-k\sigma - 0.5\sigma^2) \tag{8}$$

where k = 1.64 (5% fractile of the standardised normal distribution) is usually used. It may be helpful to consider both methods and to use the most conservative result.

This procedure may be applied for all basic variables. However, for geomechanical properties and variable loads usually other distributions apart from the normal and lognormal distribution may be more suitable.

Note that a lower acceptable reliability level can be specified by reducing β - values for probabilistic design and reducing γ - values in the partial factor method. For a material property X described by a normal distribution the partial factor γ_m may be estimated using equation

$$\gamma_m = \frac{x_k}{x_d} = \frac{\mu - k\sigma}{\mu - \alpha\beta\sigma} \tag{9}$$

which follows from general relationship (5). All the symbols used in (8) are defined above (k = 1.64 is usually used for the characteristic strength). Similar relationships between γ_m and β may be derived for lognormal or other distributions.

9.6 Structural analysis

Structural behaviour should be analysed using models that describe actual situation and state of an existing structure. Generally the structure should be analysed for ultimate limit states and serviceability limit states using basic variables and taking into account relevant deterioration processes.

All basic variables describing actions, material properties, load and model uncertainties should be considered as mentioned above. The uncertainty associated with the validity and accuracy of the models should be considered during assessment, either by adopting appropriate factors in deterministic verifications or by introducing probabilistic model factors in reliability analysis.

When an existing structure is analysed, conversion factors reflecting the influence of shape and size effect of specimens, temperature, moisture, duration-of-load effect, etc., should be taken into account. The level of knowledge about the condition of components should be also considered. This can be achieved by adjusting the assumed variability in either the load carrying capacity of the components or the dimensions of their cross sections, depending on the type of structure.

When deterioration of an existing structure is observed, the deterioration mechanisms shall be identified and a deterioration model predicting the future performance of the structure shall be determined on the basis of theoretical or experimental investigation, inspection, and experience.

9.7 Verification

Reliability verification of an existing structure shall be made using valid codes of practice, as a rule based on the limit state concept. Attention should be paid to both the ultimate and serviceability limit states. Verification may be carried out using partial safety factor or structural reliability methods with consideration of structural system and ductility of components. The reliability assessment shall be made taking into account the remaining working life of a structure, the reference period, and changes in the environment of a structure associated with an anticipated change in use.

The conclusion from the assessment shall withstand a plausibility check. In particular, discrepancies between the results of structural analysis (e.g. insufficient safety) and the real structural condition (e.g. no sign of distress or failure, satisfactory structural performance) must be explained. It should be kept in mind that many engineering models are conservative and cannot be always used directly to explain an actual situation.

The target reliability level used for verification can be taken as the level of reliability implied by acceptance criteria defined in proved and accepted design codes. The target reliability level shall be stated together with clearly defined limit state functions and specific models of the basic variables.

The target reliability level can also be established taking into account the required performance level for the structure, the reference period and possible failure consequences. In accordance with ISO 2394 [1], the performance requirements for assessment of existing structures are the same as for design of a new structure. Lower reliability targets for existing structures may be used if they can be justified on the basis of economical, social and sustainable consideration (see Annex F to ISO/CD 13822 [2]).

An adequate value of the reliability index β should be in general determined [2] considering appropriate reference period. For serviceability and fatigue the reference period equals the remaining working life, while for the ultimate limit states the reference period is in principle the same as the design working life specified for new structures (50 years for buildings). This general approach should be in specific cases supplemented by detailed consideration of the character of serviceability limit states (reversible, irreversible), fatigue (inspect, not inspect) and consequences of ultimate limit states (economic consequences, number of endangered people).

9.8 Assessment in the case of damage

For an assessment of a damaged structure the following stepwise procedure is recommended:

1) Visual inspection

It is always useful to make an initial visual inspection of the structure to get a feel for its condition. Major defects should be reasonably evident to the experienced eye. In the case of very severe damage, immediate measures (like abandonment of the structure) may be taken.

2) Explanation of observed phenomena

In order to be able to understand the present condition of the structure, one should simulate the damage or the observed behaviour, using a model of the structure and the estimated intensity of various loads or physical/chemical agencies. It is important to have available the documentation with respect to design, analysis and construction. If there is a discrepancy between calculations and observations, it might be worthwhile to look for design errors, errors in construction, etc.

3) Reliability assessment

Given the structure in its present state and given the present information, the reliability of the structure is estimated, either by means of a failure probability or by means of partial factors. Note that the model of the present structure may be different from the original model. If the reliability is sufficient (i.e. better than commonly accepted in design) one might be satisfied and no further action is required.

4) Additional information

If the reliability according to step 3 is insufficient, one may look for additional information from more advanced structural models, additional inspections and measurements or actual load assessment. The updating techniques about how to use this information have been discussed in section 9.5.

5) Final decision
If the degree of reliability is still too low, one might decide to:
- Accept the present situation for economical reasons;
- Reduce the load on the structure;
- Repair the building;
- Start demolition of the structure.

The first decision may be motivated by the fact that the cost for additional reliability is much higher for existing structure than for a new structure. This argument is sometimes used by those who claim that a higher reliability should be generally required for a new structure than for an existing one. However, if human safety is involved, economical optimisation has a limited significance.

9.9 Final report and decision

The final report on structural assessment and possible interim reports (if required) should include clear conclusions with regard to the objective of the assessment based on careful reliability assessment and cost of repair or upgrading. The report shall be concise and clear. A recommended report format is indicated in Annex G to ISO/CD 13822 [2].

If the reliability of an existing structure is sufficient, no action is required. If an assessment shows that the reliability of a structure is insufficient, appropriate interventions should be proposed. Temporary intervention may be recommended and proposed by the engineer if required immediately. The engineer should indicate a preferred solution as a logical follow-up to the whole assessment in every case.

It should be noted that the client in collaboration with the relevant authority should make the final decision on possible interventions, based on engineering assessment and recommendations. The engineer performing the assessment might have, however, the legal duty to inform the relevant authority if the client does not respond in a reasonable time.

9.10 Concluding remarks

The main principles for assessment of existing structures are:

- Currently valid codes for verification of structural reliability may be applied, historic codes valid in the period when the structure was designed, should be used only as guidance documents;
- Actual characteristics of structural material, action, geometric data and structural behaviour should be considered; the original design documentation including drawing should be used as guidance material only.

The most important step of the whole assessment procedure is evaluation of inspection data and updating of prior information concerning strength and structural reliability. It appears that a Bayesian approach can provide an effective tool.

Typically, assessment of existing structures is a cyclic process in which the first preliminary assessment is often supplemented by subsequent detailed investigations and assessment. A report on structural assessment prepared by an engineer should include a recommendation on possible intervention. However, the client in collaboration with the relevant authority should make the final decision concerning possible interventions.

9.11 References

[1] ISO 2394 (1998) *General principles on reliability of structures*. ISO, Geneva, Switzerland.
[2] FDIS 13822 (2001) *Basis for design of structures -Assessment of existing structures*. ISO, Geneva, Switzerland.
[3] ISO 12491 (1998) *Statistical methods for quality control of building materials and components*. ISO, Geneva, Switzerland.
[4] R.E. Melchers (2001) *Structural reliability analysis and prediction*. John Wiley & Sons.
[5] Ellingwood B.R. (1996) *Reliability-based condition assessment and LRFD for existing structures*. Structural Safety, 18 (2+3), 67-80.

Annex A - General flow of assessment of existing structures

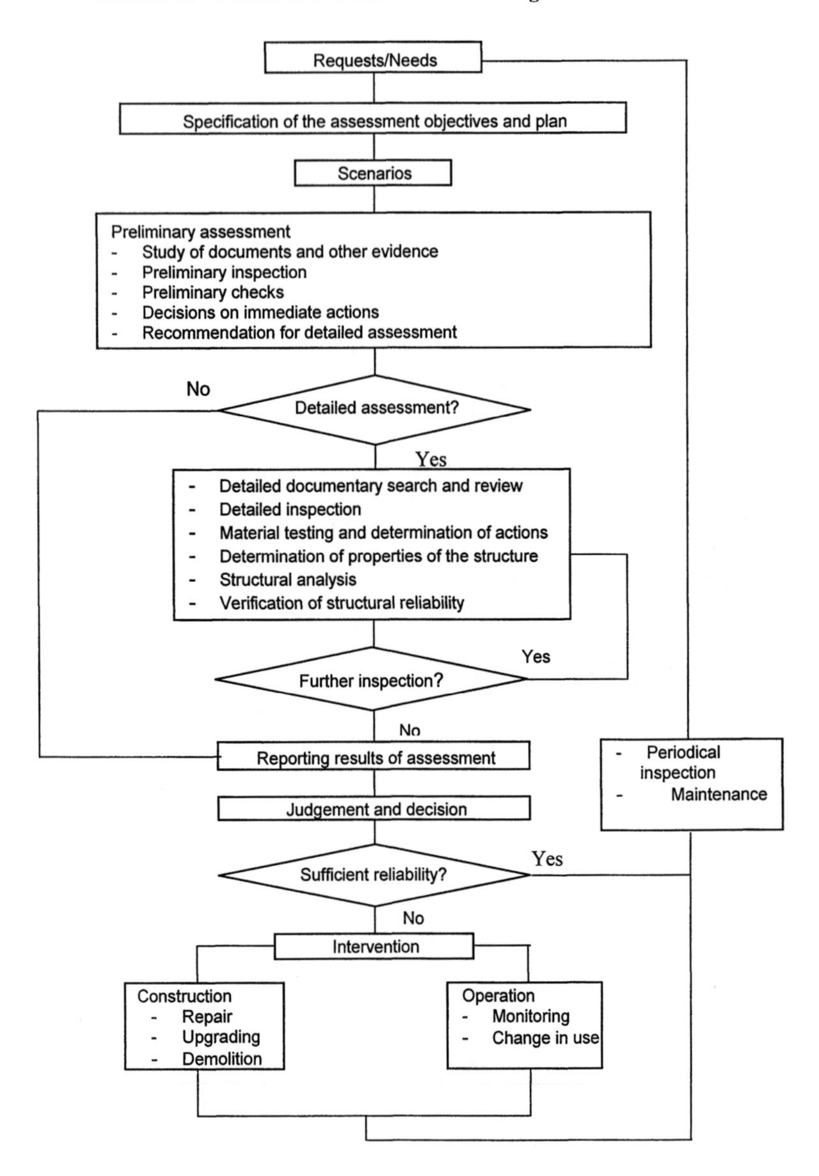

Annex B - Probability updating

This example of probability updating is adopted from [4] and [5]. Consider the limit state function G(X), where X is a vector of basic variables, and the failure F is described by the inequality G(X) < 0. If the result of an inspection of the structure I is an event described by the inequality $H > 0$ then using equation (1) in the main text the updated probability of failure P(F| I) may be written as

$$\mathrm{P}(F|I) = \mathrm{P}(\mathrm{G}(X) < 0 | H > 0) = \frac{\mathrm{P}\left(\mathrm{G}(X) < 0 \cap H > 0\right)}{\mathrm{P}(H > 0)} \tag{B.1}$$

For example consider a simply supported steel beam of the span L exposed to permanent uniform load g and variable load q. The beam has the plastic section modulus W and the steel strength f_y.

Using the partial factor method the design condition $R_d - S_d > 0$ between the design value R_d of the resistance R and design value S_d of the load effect S may be written as

$$W f_{yk} / \gamma_m - (\gamma_g\, g_k L^2/8 + \gamma_q\, q_k L^2/8) > 0 \tag{B.2}$$

where f_{yk} denotes the characteristic strength, g_k the characteristic (nominal) value of permanent load g, q_k the characteristic (nominal) value of permanent load q, γ_m partial factor of the steel strength, γ_g the partial factor of permanent load and γ_q the partial factor of variable load.

By analogy to (B.2) the limit state function G(X) follows as

$$\mathrm{G}(X) = R - S = W f_y - (\, gL^2/8 + qL^2/8) \tag{B.3}$$

where all the basic variables are generally considered as random variables described by appropriate probabilistic models.

To verify its reliability the beam has been investigated and a proof loading up to the level q_{test} is carried out. It is assumed that g_{act} is the actual value of the permanent load g. If the beam resistance is sufficient the information I obtained is described as

$$I = \{H > 0\} = \{W f_y - (g_{act}\, L^2/8 + q_{test}\, L^2/8) > 0\,\} \tag{B.4}$$

where f_y is the actual steel strength, g_{act} the actual permanent load assuming it has been determined (using non-destructive methods) reasonably accurately.

To determine the updated probability of failure P(F| I) using equation (B.1) it is necessary to assess the following two probabilities:

$$\mathrm{P}(\mathrm{G}(X) < 0|H > 0) = \mathrm{P}(W f_y - (gL^2/8 + qL^2/8) < 0 \cap W f_y - (g_{\text{act}}\, L^2/8 + q_{\text{test}}\, L^2/8) > 0) \tag{B.5}$$

$$\mathrm{P}(H > 0) = \mathrm{P}(W f_y - (g_{\text{act}}\, L^2/8 + q_{\text{test}}\, L^2/8) > 0) \tag{B.6}$$

Additional assumptions concerning then basic variables are needed. Having the results of (B.5) and (B.6) the updated probability of failure $\mathrm{P}(\mathrm{G}(X) < 0|\ H > 0)$ follows from (B.2).

Alternatively, considering results of the proof test, the probability density function $f_R(r)$ of the beam resistance $R = Wf_y$ may be truncated below the proof load as indicated in Figure B.1.

Figure B.1 Truncated effect of proof loading on structural resistance.

Obviously, the truncation of structural resistance R decreases the updated probability of structural failure defined as

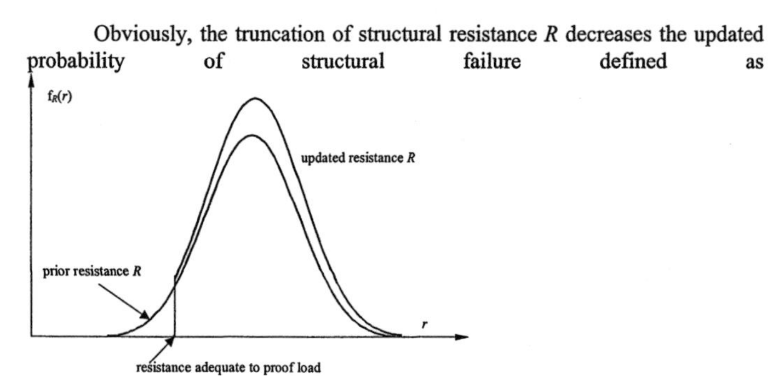

$$p_f = \mathrm{P}(R - S < 0) \tag{B.7}$$

and increase, therefore, the updated value of structural reliability.

Annex C - Bayesian method for fractile estimation

Fractiles of basic variables can be effectively updated using the Bayesian approach described in ISO 12491 [3]. This procedure is limited here to a normal variable X only for which the prior distribution function Π' (???) of ??and ? is given as

$$\Pi'(\mu,\sigma) = C\,\sigma^{-(1+\nu'+\delta(n'))}\exp\left\{-\frac{1}{2\sigma^2}\left[\nu'(s')^2+n'(\mu-m')^2\right]\right\} \tag{C.1}$$

where C is the normalising constant, ??(n') = 0 for n' = 0 and ??(n') = 1 otherwise. The prior parameters m', s', n', ?' are parameters asymptotically given as

$$\mathrm{E}(?) = m',\ \mathrm{E}(?) = s',\ V(?) = \frac{s'}{m'\sqrt{n'}}\ ,\ V(\sigma) = \frac{1}{\sqrt{2\nu'}} \tag{C.2}$$

while the parameters n' and ?' are independent and may be chosen arbitrarily (it does not hold that ?' = n' – 1). In equation (C.2) E(.) denotes the expectation and V(.) the coefficient of variation of the variable in brackets. Equations (C.2) may be used to make estimates for unknown parameters n' and ?' provided the values V(?) and V(?) are estimated using experimental data or available experience.

The posterior distribution function ?"(?,?) of ? and ? is of the same type as the prior distribution function, but with parameters m'', s'', n'' and ?", given as

$$\begin{aligned} &n'' = n' + n \\ &?'' = ?' + ?? + ?\,(n') \\ &m''n'' = n'm' + nm \\ &?''(s'')^2 + n''(m'')^2 = ?'(s')^2 + n'(m')^2 + ?s^2 + nm^2 \end{aligned} \tag{C.3}$$

where m and s are the sample mean and standard deviation, n is the size of the observed sample and ??= $n-1$. The predictive value $x_{p,\mathrm{pred}}$ of a fractile x_p is then

$$x_{p,\mathrm{Bayes}} = m'' + t_p s''\sqrt{1+1/n''} \tag{C.4}$$

where t_p is the fractile of the t-distribution (see Table C.1) with ?" degrees of freedom. If no prior information is available, then n'= ?'= 0 and the characteristics m'', n'', s'', ?" equal the sample characteristics m, n, s, ?. Then equation (C.4) formally reduces to so called prediction estimates of the fractile given as

$$x_{p,\mathrm{pred}} = m + t_p s\sqrt{1+1/n} \tag{C.5}$$

where t_p denotes again the fractile of the t-distribution (Table C.1) with ? degrees of freedom. Furthermore, if the standard deviation ? is known (from the past experience), then ???= ∞ and s shall be replaced by ?.

Table C.1 - Fractiles – t_p of the t-distribution with ???degrees of freedom

?	$1-p$					?	$1-p$				
	0.90	0.95	0.975	0.99	0.995		0.90	0.95	0.975	0.99	0.995
3	1.64	2.35	3.18	4.54	5.84	12	1.36	1.78	2.18	2.68	3.06
4	1.53	2.13	2.78	3.75	4.60	14	1.35	1.76	2.14	2.62	2.98
5	1.48	2.02	2.57	3.37	4.03	16	1.34	1.75	2.12	2.58	2.92
6	1.44	1.94	2.45	3.14	3.71	18	1.33	1.73	2.10	2.55	2.88
7	1.42	1.89	2.36	3.00	3.50	20	1.32	1.72	2.09	2.53	2.85
8	1.40	1.86	2.31	2.90	3.36	25	1.32	1.71	2.06	2.49	2.79
9	1.38	1.83	2.26	2.82	3.25	30	1.31	1.70	2.04	2.46	2.75
10	1.37	1.81	2.23	2.76	3.17	∞	1.28	1.64	1.96	2.33	2.58

Example

A sample of $n = 5$ concrete strength measurements having the mean $m = 29.2$ MPa and standard deviation $s = 4.6$ MPa is to be used to assess the characteristic value of the concrete strength $f_{ck} = x_p$, where $p = 0.05$. If no prior information is available, then n'= ?'= 0 and the characteristics m'', n'', s'', ?" equal the sample characteristics m, n, s, ?. The predictive value of x_p then follows from (B.5) as

$$x_{p,\text{pred}} = 29.2 - 2.13 \times \sqrt{\frac{1}{5}+1} \times 4.6 = 18.5\,\text{MPa}$$

where the value $t_p = -2.13$ is taken from Table C.1 for $1-p = 0.95$ and $\nu = 5-1 = 4$. When information from previous production is available the Bayesian approach can be effectively used. Assume the following prior information

$$m' = 30.1\ \text{MPa},\ V(m') = 0.50,\ s' = 4.4\ \text{MPa},\ V(s') = 0.28$$

It follows from equation (C.2)

$$n' = \left(\frac{4.6}{30.1}\frac{1}{0.50}\right)^2 < 1\ ,\ \nu' = \frac{1}{2}\frac{1}{0.28^2} \approx 6$$

The following characteristics are therefore considered: $n' = 0$ and ?' = 6. Taking into account that $\nu = n - 1 = 4$, equations (B.3) yield

$$n'' = 5,\ \nu'' = 10,\ m'' = 29.2\,\text{MPa},\ s'' = 4.5\,\text{MPa}$$

and finally it follows from equation (C.4)

$$x_{p,\text{Bayes}} = 29.2 - 1{,}81 \times \sqrt{\frac{1}{5}+1} \times 4.5 = 20.3\,\text{MPa}$$

where the value $t_p = -1.81$ is taken from Table C.1 for $1-p = 0.95$ and $\nu = 10$.

In this example the resulting characteristic strength is greater (by about 10 %) than the value obtained by prediction method without using prior information. Thus, when previous information is available the Bayesian approach may improve (not always) the fractile estimate, particularly in the case of a great variance of the variable. In any case, however, due caution should be paid to the origin of the prior information with regard to the nature of considered variable.

Chapter 10

Quality Management of structural design

By:
Edmund Booth, Consulting Engineer, London

10.1 Purpose and scope of this chapter

The purpose of this chapter is to provide an introduction to quality management and Quality Assurance (QA) tools used during the structural design of buildings. The chapter is based on the definitions and procedures set out in the international standards ISO 8402, 9001 and related documents and makes special reference to structural safety issues. The scope includes all heights of building, with the final section discussing some special considerations for tall buildings. QA during construction and maintenance is not included, since it is covered in Chapter 11 of this monograph. QA is treated in the specific context of the management systems associated with the attainment of quality in structural design (as defined in section 10.2.2 below), rather than the broader context of achieving excellence in structural design.

10.2 Overview of Quality Management in structural design

10.2.1 Background

Quality assurance (QA) has its origins in manufacturing industry, where the quality of a physical product supplied to a customer was the issue at stake. Much of the material on QA is related primarily to the needs of manufacturing industry. QA can still bring significant benefits in the rather different context of design consultancy, where the product is a design or a report rather than a manufactured object; this different context must however be borne in mind when drawing up a quality management system.

QA has assumed increasing importance in the last decade for structural design consultancies (ie organisations carrying out structural engineering design work), to the point where a majority of large building design projects carried out by UK practices, and nearly all those for industrial projects, are covered by a formal quality assurance system. A similar situation applies in many other parts of the world.

10.2.2Definitions

Internationally agreed definitions relating to QA are given in ISO 8402 (1994). Some important definitions are given in Table 10.1.

Term	ISO 8402 (1994) Definition	Comments
Quality	The totality of characteristics of an *entity* that bear on its ability to satisfy stated and implied needs.	An *entity* is defined as a process, product or organisation; in the current context, it is the structural design of a building, embodied in a set of drawings, specifications and so on.
Quality System	Organizational structure, procedures, processes and resources needed to implement quality management	
Quality Management	All activities of the overall management function that determine the quality policy, objectives and responsibilities, and implement them by means of quality planning, quality control, quality assurance and quality improvement, within the quality system.	
Quality Assurance	All the planned and systematic activities and functions implemented within the Quality System and demonstrated as needed to provide adequate confidence that an *entity* will fulfil requirements for quality.	QA, like quality management, is a *management,* rather than a *technical* procedure. Its concern is checking that technical (and other) measures have been carried out which are needed to ensure that an *entity* (in this case, design of a building structure) meets the safety and other requirements set out for it. The technical measures themselves (calculation checks, physical testing etc) are not part of QA, although they must be carried out to satisfy QA requirements.
Quality Control	Operational techniques and activities that are used to fulfil requirements for quality	Quality Control involves the use of (mainly technical) tools to examine a product for compliance with its stated requirements for quality. In the design context, arithmetical checks on calculations and checking conformity of drawings to draughting standards are examples of Quality Control.

Table 10.1: Definitions of some QA terms

Further discussion is needed in the definition of the 'product' represented by the structural design of a building. There are two aspects of the product which QA needs to cover. Firstly, there is the quality of the physical representation of the design - the draughting standards for drawings, setting out of calculations, layouts of reports and so on. Secondly (and much more importantly) there are the inherent characteristics of the structure implied by the design, such as its safety, serviceability and so on. Note however that the ultimate physical reality of the building structure is not part of the product of structural design; QA aspects for this ultimate stage are discussed in Chapter 11 of this Monograph. The definition of 'product' in the context of design consultancy is further discussed in section 3.11 of CIRIA Special Publication 84 (Barber, 1992).

10.2.3 Safety and QA

As is apparent from the definitions in Table 10.1, QA is concerned with the attainment of all those features of a design which bear on its ability to satisfy a client's needs. In addition to these client needs, the wider needs of society (as for example set out in regulatory documents) must also be addressed. As a consequence, an engineering design must satisfy a wide range of different requirements, including the following (based on Matousek, 1992).

- Serviceability
- Safety
- Environmental compatibility and sustainability
- Durability and reliability
- Economy and efficiency
- Constructability
- Aesthetic appeal

Thus it must be recognised that safety is only one of a number of aspects of 'quality', though it is clearly a very important one.

It should also be appreciated that the Quality Assurance of the design process provides solely a management framework within which an appropriate level of structural safety can be achieved in a design; the safety itself is achieved only by the application of well founded engineering principles.

10.2.4 The aim, costs/benefits, dangers and limitations of QA in the structural design process

10.2.4.1 Aim of QA

The aim of QA is to produce a product which is "right first time, every time" by eliminating errors and omissions and ensuring a product that fits the customer's requirements.

10.2.4.2 Benefits & costs of QA

The benefits claimed for a successful quality management system include the following.

- Reduction in errors and abortive work
- Better customer satisfaction through a product (in the present context, building design) that is better suited to the client's needs
- More efficient working through standardised and well recorded procedures
- Continuous improvements in standards through the QA review process, which as discussed later is a central feature of QA.

A well documented record of the design process, and the rationale behind it, should result from the successful application of QA. The benefits ensue from the following essential characteristics of the QA process, which are referred to again in later sections.

- QA necessitates a well recorded process which can be audited, and hence checked
- The QA documentation records the things that have worked well, and allows them to be applied in future.
- Equally importantly, when things go wrong the documentation can be reviewed to discover the weaknesses and correct them in future.

There are also a number of benefits which apply after construction of the building, as follows.

- The documentation will be helpful where structural changes or demolition need to be carried out after original construction has been completed.
- Well documented records of any maintenance or monitoring requirements should also ensue.
- In the event of dispute or litigation, the existence of 'objective evidence' provided by the quality management system should assist the designer's case.

The benefits listed above can also lead to associated financial benefits, though for building projects (particularly tall ones) which are generally unique, the financial gain is very hard to quantify. There are offsetting costs, associated with the initial setting up and training necessitated by a new quality management system and with the continued running of the system. General opinion within the UK consulting engineering community is that although the initial costs are substantial, the ongoing costs are much less and overall the costs are justified by the benefits.

10.2.4.3 Dangers of QA

There is a danger that QA can degenerate into a costly self-perpetuating paper chase. This can be reduced by adhering to the following fundamental principle of a good quality management system, slightly adapted from ISO 8402 section 3.6:

> 'The quality system should [only] be as comprehensive as is needed to meet the quality objectives'. (The word *only* has been added to ISO 8402).

However in the past, particularly where systems have been too prescriptive, this danger has not been avoided. QA has been used merely to tick off what has to be done rather than deal thoughtfully with the issues, or has focused narrowly on the audit issues rather than addressing the broader, underlying ones. QA has received a bad reputation in some quarters as a result, but can bring real and substantial benefits, provided the pitfalls referred to are avoided.

Another danger of QA is that it may lead to a false sense of security - the assumption that because a quality management system is in place, all safety and other 'quality' issues are automatically taken care of. Properly instituted, a quality management system should lead to the opposite situation, by making sure that all parties to a design are made aware of their responsibilities for 'quality' through the planning and review process that is central to QA. However, this is only likely to come about if those with overall responsibility for the design process take active steps to ensure that the requirements of the quality management system are implemented.

10.2.4.4 Limitations of QA

There are limitations in the application of QA to structural design consultancy (as elsewhere), and a quality management system only provides a framework within which talented, well qualified engineers can produce designs of excellence. By itself, the quality management system cannot ensure excellence; it gives assurance of the framework for performing the work, rather than the content of the work itself.

These limitations apply just as much too structural safety as to the other aspects of design covered by QA. Thus QA is not a panacea; there have been instances of serious structural failure occurring where quality management systems were nominally in place. Nevertheless, in the right environment QA does have the potential to facilitate the achievement of structural safety.

10.2.5 Basic concepts and requirements of a Quality Management System

This section introduces the most important elements of QA and quality management in structural design consultancy. More specific details are given in Section 0.

10.2.5.1Fundamental concepts

A fundamental concept of successful QA is the institution of a cyclic process, as follows.

First you	PLAN	WHAT YOU DO (IN WRITING)
Then you	DO	WHAT YOU PLANNED
You must	RECORD	THAT YOU DID IT
Then you	REVIEW	WHAT YOU DID
So that you can	IMPROVE	WHAT YOU DO NEXT TIME

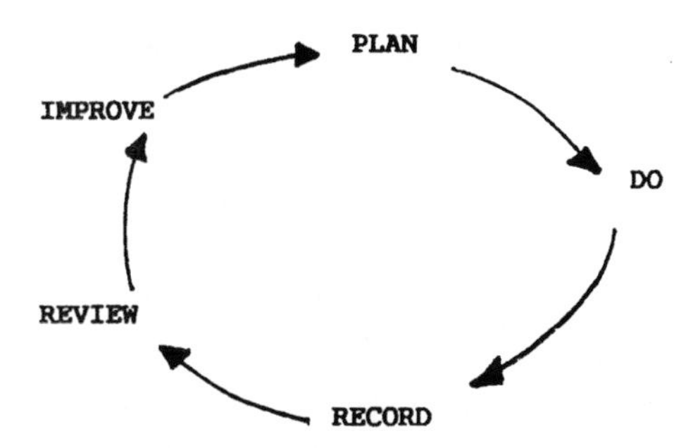

Figure 10.1: Cyclic review in QA

Another fundamental concept of QA is that everyone accepts responsibility for their own work. The Quality Management System provides the framework within which individual responsibility can be encouraged and effectively discharged to complement the system of independent checks for confirming that requirements for quality are being met.

10.2.5.2 Essential requirements

Three essential requirements of a Quality Management System applying to a structural design consultancy are high level *commitment*, appropriate *documentation* and *audits*. These three requirements are now discussed in turn.

a) Commitment at a high level

It is essential that there is a commitment to implementing and adhering to an organisation's Quality Management System at the highest level in the organisation. A written policy statement to this effect, signed by the senior member of staff responsible for QA, is often contained in an organisation's Quality Management System documentation.

b) Quality Management System documentation

Several levels of Quality Management System documentation are required. At the most general level, organisation wide standard procedures and methods need to be defined, setting out for example the format for project plans, and recording procedures such as minutes of meetings or document transmittals. In large organisations, these general procedures may need to be fleshed out for the specific purposes of groups or sub-groups within the organisation; hence group Quality Plans may be needed. An important function of these general elements of the Quality Management System documentation is to ensure that the recording of actions is carried out in a well controlled and standard fashion.

The most specific piece of documentation, which forms an essential part of the planning process referred to in section 0, is the project specific quality plan, which sets out the procedures, goals and requirements for a particular project; the project quality plan is discussed in section 3.2.

c) Audits

Audits are designed to check whether the Quality Management System is being adhered to. Audits require objective evidence, hence the crucial role of recording after action, as referred to in section 0.

10.2.6 International Standards for Quality Management Systems

A number of International Standards Organisation (ISO) standards on quality management systems have been published. They are based on British standards, and have also been published as European Norms (ENs). United States and New Zealand QA standards are also based on the ISO documents. The ISO standards are general documents which provide a basic framework from which procedures appropriate to particular organisations must be developed. The two ISO standards most relevant to QA in structural design consultancy are as follows.

ISO 9000:2000 Quality management and quality assurance standards

Part 1: Guidelines for selection and use.

Part 2: Generic guidelines for the application of ISO 9001, 9002 & 9003.

Part 3: Guidelines for the application of ISO 9001 to the development, supply and maintenance of software.

IS0 9001:2000 Quality Systems - Specification for design development, production, installation and servicing.

Sixteen other ISO standards are referenced in Annex E of ISO 9000 Part 1. Of particular relevance to structural design consultancy are the following:

ISO 10011:1990: Guidelines for auditing quality systems, Parts 1 to 3.

ISO 8402:1994 Quality management and quality assurance - vocabulary.

ISO 9000 and 9001 were written primarily for organisations that carry out both design and production of a product. The standards' application therefore needs interpretation and adaptation to design consultancies, because often their brief may not include site supervision or may be restricted to scheme design only. However, if the 'product' of structural design is literally interpreted as a set of drawings and calculations, and the QA process is taken as being confined to covering just the physical characteristics of that output (draughting standards and so on) little will be gained with respect to structural safety. Therefore, as suggested in section 2.2, the 'product' must also be taken to include the inherent characteristics of the structure implied by the design, in terms of its safety, serviceability and so on. This is essential if the QA process is to deliver real benefits. With this broader definition of 'product', ISO 9000 and 9001 have been successfully adopted as the basis of QA systems capable of delivering a framework for producing structural designs which properly address safety, as well as other key issues.

10.2.7 Components of QA for a specific project

ISO 9001 Clauses 4.3 & 4.4 defines ten components relating to QA of the design process for a specific project, which need to be carried out within the context of an organisation's Quality Management System. These ten components are listed below. The implementation of QA in a structural design project is discussed in more detail in section 0.

i) Contract review
This is carried out to check that the client's requirements are fully and unambiguously defined and recorded, and that the structural design consultant has the capability to meet those requirements.

ii) Establishment of documented procedures &
iii) Design and development planning
As discussed in Section 0, recording and planning are central to the QA process.

iv) Organizational and technical interfaces
Interfaces between different groups having an input into the design process need to be defined so that the necessary information flows across them and also so that the division of design responsibility is clear. Errors often arise at interfaces.

v) Design input
The design input requirements identified in the contract review, with applicable statutory and regulatory requirements, should be documented and appropriately distributed.

vi) Design output
Design output (drawings, calculations, specifications, reports etc.) should be recorded in a form that can be verified and validated against the design input requirements. The quality management system needs to check the following:

a) that the verification and validation against requirements has taken place (though technical review is the responsibility of the design team, rather than the QA process per se)
b) that the design output contains or makes reference to acceptance criteria, such as client's specifications and regulatory requirements
c) that there is identification of those characteristics of the design that are crucial to safe and proper functioning of the building.

vii) Design review
The QA function is to ensure that the design reviews take place, are appropriately attended and properly recorded, and that the actions arising from the review are followed through.

viii) Design verification &
ix) Design validation
Verification relates to the checking and approval process, while validation, which normally follows successful verification, relates to confirmation that the final design conforms to the user needs and requirements.

x) Design changes
The QA activity is to check that design changes have been properly identified, recorded and approved before implementation, and that all necessary parties have been made aware of the changes.

10.3 Quality Assurance during structural design

10.3.1 Introduction

The following sections provide a brief introduction to the implementation of QA during a structural design project. Sections 3.2 to 3.5 discuss the various activities carried out during structural design which need to be addressed by QA, while section 3.6 relates these activities to the requirements of ISO 9001 referred to in section 2.7.

Of course many elements of quality management systems, such as conducting appropriate Contract and Project Reviews, were standard good practice long before the adoption of formal quality management systems by structural engineering consultants. The benefits of introducing a formal Quality Management System are as follows.

- Provision of a framework for planning and recording the design process.
- Integration of that framework into the totality of the checking systems of a project.
- Helping to ensure uniformity of practice between projects.
- Making the process clear and understandable to those not directly involved.
- Ensuring that the division of design responsibilities is clear.
- Ensuring that complete and up to date information is distributed to those who need it.
- Ensuring that experience is recorded and hence able to be passed on; this includes lessons learned from mistakes and new solutions to difficult problems.

Further discussion of QA in the service industry is provided by Stebbing (1990), while ACE (1988), ACEA (1990) and ASCE (1988) also provide useful material.

10.3.2 The Project Quality Plan

The Project Quality Plan is the most important QA document produced during the planning stage of a project. Its scope and detail should be tailored to the specific requirements of the project, bearing in mind the maxim referred to previously "The quality system should only be as comprehensive as is needed to meet the quality objectives". Typically, the Project Quality Plan may contain the following elements.

- Identification of key personnel in the client, supplier and subcontractor organisations, designating responsibilities
- Client's brief
- Programme of activities, indicating key milestone dates
- Administrative procedures, such as special administrative arrangements
- Scope, frequency and timing of Project Reviews, and identification of the reviewers
- Identification of technical procedures, such as references to calculation plans, relevant codes of practice and regulatory requirements. These include regulatory requirements for safety; in the UK, these would include the Construction (Design & Management) Regulations (1994), intended to enhance health and safety in the construction industry and the Building Regulations, which set out minimum performance standards for buildings, including safety standards.
- Identification of checking procedures for technical calculations, drawings, software etc., to include the scope and rigour of the checking procedures.

10.3.3 Contract and Project Reviews

Contract reviews are required before acceptance of the contract for a project, to ensure that the client's requirements are clear and unambiguous, and that the structural design consultancy has the capability for meeting those requirements.

Project Reviews are intended to take an overview of a project at key stages after inception, to ensure that the client's requirements are being met in the most satisfactory way. Their purpose is quite distinct from the audits discussed in section 0. Staffing, programming and cost matters may also need to be addressed, as do any issues arising from previous project or contract reviews.

Issues related to safety include ensuring that no potential failure modes, unusual loading patterns, or loading types have been neglected. The general issue of whether adequate robustness and redundancy have been provided, which is so crucial to structural safety, should also be addressed. Applicable safety regulations need to be complied with. It is also important that the wider implications of late changes in design are properly considered; failure to do so has in the past given rise to structural failure. Design changes should be subject to the same verification procedures as the original design. Overall, project reviews can make a crucial contribution to ensuring that structural safety is adequately addressed.

Project reviews should involve personnel from the organisation not directly engaged on the project, particularly those with experience of similar projects, and may also include personnel with specific expertise from outside the organisation.

Such reviews are required at significant milestones in a project. At minimum, this will be upon acceptance of a client's brief, and immediately before completion of the project. On large projects, reviews may also be conducted at the completion of other stages, such as scheme design, design development, detailed design, and so on. Reviews are likely to be necessary at 'handover' stages where significant input begins from new groups, either from the same organisation or from outside it. This is to ensure that issues of interface between design groups are properly addressed.

As previously explained, the Quality Management System is concerned with checking that Project Reviews are properly carried out and recorded; it does not become involved in the technical details of such checks *per se*.

10.3.4 Detailed checking procedures

While Project Reviews are concerned with broad issues, the minutiae of a project's design also need checking. The level of detail of checking will depend on the importance and complexity of the project and should be set out in the Project Quality Plan. A number of different methods such as spot checks and parallel calculations may be used. Calculations, computer programs, computer output and drawings may all need to be checked. Special consideration is needed for cases where intermediate steps in a process are not apparent. An example is the output of a spreadsheet calculation where the input and output may be presented without the intermediate steps (for example, arithmetical manipulation) being apparent. The same applies more generally to the output of specialist software.

The scope and scale of the checking process (for example, whether spot checking, 100% checking, independent parallel calculation, or whatever) will depend on the nature of the element being checked, and the consequences of error (particularly in a safety context). It is fundamental, however, that the designer carrying out a particular procedure is responsible for its correctness, irrespective of the number of checks to be carried out by others, and self-checking is essential in all cases. Again, this is particularly important in the context of checking for safety.

Arithmetical checks may (and not infrequently will!) reveal errors, which may have implications for safety. However, a 100% arithmetically correct calculation which is flawed in the way it models the physical world can still produce an unsafe result; simple arithmetical checks will only reveal such flaws if the checker uses judgement, experience and vigilance and carefully examines the underlying assumptions. Independent parallel checks can be more successful in this respect. However, life threatening structural failures due to structural design errors has generally been ones that could have been detected or prevented at design review level, rather than by detailed checking.

Once again, the procedures described above are not unique to a quality management system, which merely provides a framework in which they can take place and be recorded.

10.3.5 Quality assurance of the dissemination of structural design information

Timely supply of properly checked documentation to the appropriate personnel, and avoiding the use of superseded documentation, has long been a concern of the construction industry. QA provides a rational and structured framework within which tried and tested methods can operate. Drawings, calculations, design reports and source documents (such as codes of practice and standard specifications) may all be involved.

When electronic means of documentation transmission are used, the traditional methods associated with paper documents need to be adapted. Document control software can provide comprehensive systems of control where all documents are stored in electronic format (Hayward, 1998). 'Project websites' (also known as extranets or project collaboration systems) can establish a database of information for sharing between all parties in the design, construction and subsequent maintenance & operation of a building. These websites have the potential to improve assurance, compared with paper based systems, particularly where a number of different organisations are involved. The resulting ease with which information is shared between teams should lessen problems arising at the interfaces between design responsibilities. The project website will also allow identification of who has had access to information. This should assist the audit trail (see section 4.0) that is part of the QA process in checking that the necessary information has reached the right places. Clearly this implies that the matrix of distribution and access to documents has been defined in the project quality plan.

10.3.6 Relationship to ISO 9001

Table 10.2 relates the QA activities discussed in sections 3.2 to 3.5 to the elements of QA for design set out in IS0 9001, as discussed in section 2.7.

Contract review and design control items listed by ISO 9001	Project Quality Plan	Contract review	Project reviews	Detailed checking	Design information dissemination
i) Contract review		□			
ii) Documented procedures	□				
iii) Design & development planning	□				
iv) Organizational and technical interfaces	□		□		
v) Design input	□				
vi) Design output					□
vii) Design review			□		
viii) Design verification			□	□	
ix) Design validation			□ (Note 1)		
x) Design changes					□

Table 10.2: Matrix of QA activities and ISO 9001 requirements

Note 1: The final project review before issue of contract documents is likely to form the main means of design validation for the structural design of a building. Commissioning of the finished construction may more rarely fall within the structural design consultant's brief.

10.4 Auditing Quality Management of the structural design process

ISO 10011 defines audits as providing a "systematic and independent examination to determine whether quality activities and related results comply with planned arrangements and whether these arrangements are implemented effectively and are suitable to achieve objectives". They are thus quite distinct from the review and checking procedures discussed in sections 0 and 3.4. The audit procedure lies at the heart of a quality management system; for structural design consultancy, it involves checking that systems are in place to carry out the ten components of QA listed in section 0, that they are properly implemented in individual projects and that non-compliance with any of the requirements laid down by the Quality Management System are followed through and resolved. However, the technical procedures for review and checking do not themselves form part of the Quality Management System, which is a management procedure.

Audits can be used to assess the overall operation of the Quality Management System for the whole organisation or one of its sub-groups. They can also assess the specific operation of QA within a project, generally using the Project Quality Plan as a benchmark. Various types of audit are possible, depending on whether the audits are carried out internally or externally to the organisation.

Audits will typically consider the following aspects.

Procedure Assessments.
Are the adopted operating procedures adequate for the purpose? Do they adequately define and control particular operations?

Operation Assessments
Can it be seen that what is being done is in accordance with the written system?

Product Assessment
Does the design satisfy the Client's requirement, as set out in the Brief?

Document Assessment
Does Objective Evidence exist that the Quality Management System is being properly followed? Is there an Audit Trail which can be traced through each stage of the delivery of the service?

10.5 Special considerations for tall buildings

The overall Quality Management system set up within an organisation and its sub-groups should be sufficiently flexible that it can accommodate a large variety of different types of project. Most of what has been written so far applies as much to low rise as to tall buildings. However, the Project Quality Plan for a 100 story skyscraper would be very different from that for a private residential dwelling (were such a plan to exist) because of the size and complexity of the former. Examples of what these differences might include are now given.

By definition, tall buildings are major projects, both economically and also in potential for affecting the life safety of a large number of people. Therefore, in most cases, more comprehensive QA measures are appropriate than is usually the case for low rise buildings. In particular, there will be an increased frequency of project reviews, which will be required at concept, scheme and detailed design stages (perhaps several times in each), and more extensive and more independent checking of calculations, drawings and so on.

Tall buildings may involve unusual engineering problems for which precedents are less available than usual. For example, the response to wind and perhaps seismic loading is likely to require special techniques. Novel engineering solutions, such as active or passive control devices, may be proposed. This has implications for the planned review and checking process, and also for the project calculation plan set out in the Project Quality Plan.

The collapse of the World Trade Centre (WTC) towers in New York in September 2001 will inevitably colour the thinking of designers of future tall buildings, particularly very tall ones. The robustness of important structures and their ability to resist progressive collapse have always been important, but these aspects are bound to feature even more prominently in the design process and project reviews will need to consider carefully the type of accident scenario that needs to be accounted for. Fire safety and means of egress, although strictly beyond the scope of this monograph, are also vital aspects that need to be considered. QA in itself has nothing directly to say on these topics; however, it can properly ensure that personnel with the right qualifications and experience take part in the project reviews and elsewhere in the design process, and that relevant guidance on the safety of tall buildings prepared in the light of the WTC disaster forms part of the project input documentation.

The size of project, number of personnel and perhaps number of different subcontractors, and the accompanying volume of paperwork generated, may pose special documentation control problems which need to be addressed in the Project Quality Plan. Clear lines of communication and procedures for document transmission become vitally important. Equally, clear division and allocation of responsibility is very necessary, and these need to be addressed by project reviews at appropriate milestones in the project.

10.6 Acknowledgements

This chapter was prepared with the assistance of Richard Henderson and John Ward of Ove Arup & Partners, and their essential contribution is gratefully acknowledged. Some of the material is adapted from that prepared by the Ove Arup Partnership for internal use, and this is also gratefully acknowledged. Extensive and valuable discussions were held with John Barber of Kings College, London, and helpful comments were obtained from David Elms of University of Canterbury, New Zealand, Keith Seago of Ove Arup & Partners, Charles Botford of BIW Technologies Limited and David Blockley of University of Bristol.

References

ACE 1988.
QUALITY ASSURANCE. Guidance Note with Synopses for Quality Manual and Operational Procedures. The Association of Consulting Engineers, London.

ACEA 1990
Quality management: a guide for professional design practice. Association of Consulting Engineers of Australia, North Sydney.

ASCE 1988
Quality in the constructed project: a guideline for owners, designers and constructors. American Society of Civil Engineers, New York.

Barber J (1992)
CIRIA Special Publication 84: Quality Management in construction - contractual aspects. Construction Industry Research and Information Association, London.

Hayward D. 1998
DOCUMENTATION MANAGEMENT - OUT OF PAPER. New Civil Engineer Special Supplement, February, II-V. Emap Construct, London.

Matousek M. 1992.
QUALITY ASSURANCE. In: Engineering Safety (ed. D Blockley). McGraw-Hill Book Company, London, 72-88.

Stebbing L. 1990.
QUALITY MANAGEMENT IN THE SERVICE INDUSTRY. Ellis Horwood, Chichester, UK.

Chapter 11

Quality Management in Construction

By:
Shunsuke Sugano, Hiroshima University, Higashi-hiroshima 739-8527, Japan

11.1 Purpose and Scope of This Chapter

This chapter describes the process of quality management in construction of buildings. It is hard to comprehensively describe quality control in construction of buildings and to provide guidelines for achieving quality control in construction because each country has established its own building production systems based on its historical background, markets and traditions. Therefore, the concept of quality management in building construction and case studies described in this Chapter pertains to construction practices in Japan. Moreover, emphasis is placed on the construction of high-rise concrete buildings, in which more technical development has been required and the quality control has been more difficult to achieve in comparison with high-rise steel buildings.

11.2 Overview of Quality Management in Construction

11.2.1 Concept of quality management in building production

The quality of design of a building, consistent with design documents and specifications, can be said to be "aimed quality" while the quality achieved in the construction of a building can be said to be "confirmed quality". Activities that satisfy quality requirement items may be defined as "quality management" while activities that give confidence in the quality achieved to building owners and users may be defined as "quality assurance". Quality assurance incorporates quality management. Furthermore, comprehensive quality control (CQC) and total quality management (TQM) are effective approaches to achieve quality management and quality assurance.

In the construction industry, many enterprises are requiring compliance with the international standards ISO 9000 for quality assurance and quality management. The ISO 9000 standards classify the items required for the system to create the quality (quality system) into 20 items described as follows.

1) Manager's responsibility.
2) Quality system.
3) Identification of contents of contract.
4) Design and supervision.
5) Charge of documents and data.
6) Purchase.
7) Charge of client supplies.
8) Distinction of products and traceability.
9) Management of work progress.
10) Inspection and examination.
11) Inspection, measurement and test equipment.
12) Condition of inspection and examination.
13) Management of un-conformed products.
14) Correction and prevention treatments.
15) Handling, wrapping, storage and delivery.
16) Charge of quality control record.
17) Interior quality inspection.
18) Education and training.
19) Service.
20) Statistical methods.

The Japanese guidelines for quality control in concrete construction, proposed by the Architectural Institute of Japan in 1999, describe the quality control process should be implemented for many of the items corresponding to those above.

11.2.2 Building production system and quality management

The building production is generally achieved by the "design-construction segregate" system where the design and the construction are performed independently by separate enterprises and organizations in charge. Large general contractors often produce buildings through the "design-build system" where all the processes in building production are performed by one company. Comprehensive quality control may be easily implemented within this system because all the operations for building production are achieved by one company. In addition, many of these general contractors have technical institutes for research and development supported by their own resources. There are many research activities in those institutes that produce new type of buildings and structures and develop construction technology as well as to support design and construction practices. Such research and development activities contribute greatly to the design and construction quality.

Quality management is achieved in both stages of design and construction. There is the idea that "a post-process is a client". For example, a designer considers the general contractor to be an important client; in turn, the general contractor considers a specialist contractor to be an important client. This way of thinking is important for quality management in overall building production because it is essential to understand and communicate with the "post process" and to achieve each process so that difficulties and waste do not accumulate. The history of quality management shows that this way of thinking is connected with the rationality and the economy of the resulting building construction product.

11.2.3 Special considerations to high-rise buildings

As a building becomes taller, difficulties in design and construction increase and it becomes difficult to assure the required building quality. The design of a building beyond 60m high must be subjected to an independent review and appraisal by a code authority like the Building Center of Japan (BCJ). Furthermore, when a special method of construction or material is used, the construction also must be subjected to an appraisal by the BCJ. In the case of high-rise reinforced concrete buildings, high strength concrete is handled as a special material beyond the usual range, and always becomes a subject for the appraisal. In particular, it must be verified that the required quality of concrete can be produced and constructed with a high degree of confidence.

11.3 Quality of Building and Quality Assurance during Construction

11.3.1 Quality of building

A building protects individuals, enterprises and national life and property. Furthermore, the building represents the culture of that period for the future. Therefore, it can be said the building is not a mere "product" or "merchandise" but a "work". The building must maintain freshness in capability, economy, artistry and social relevance over a long period of time. Such a building will be a source of satisfaction to the owner, occupants and the public at large.

11.3.2 Creation of quality

Quality must be initiated by fully understanding the client's needs and wishes in conceptualizing the previously described building. The client's needs are put together as the "aimed quality". They are reflected in the design drawings and specifications, and they are faithfully created during the construction stage on site as the "confirmed quality". If the basic functions that a building must possess cannot be achieved, the client will be disappointed and possibly angry. A series of activities are conducted to assure the quality: setting the basic function in the design stage, creating the "work" in the construction stage and servicing after the completion. Problems encountered at each stage are fed back to the previous stages of the process to prevent their recurrence. It is important to secure the previously described ordinary quality in the really attractive "work". Continuous comprehensive quality management activity is necessary to realize this goal.

11.3.3 Quality assurance in individual work

The quality assurance scheme for individual work for a building is divided into several steps, and the policy of the quality assurance is developed into all stages, as shown in Fig. 1. In the design stage, the fundamental functions and performance requirements are defined, and a three-step design review is carried out. Those ingredients are reflected in the building construction process. The output information in each step is summarized in each division in charge and analyzed. Each process is improved by giving the results to the upper stage of the individual activity for quality assurance.

Construction management is performed with the cooperation of other companies involved and the "confirmed quality" is made in the steps of Q7 through Q11 in the construction stage. Feedback to the previous step is analyzed, improved and standardized at the company level. Thus the quality assurance is made uniform.

11.3.4 Technology development

The technology development becomes the motive power for new buildings and products in response to the requirement of the time. The technology development is conducted from two sides. One side is the development of technology in response to social needs. Another side is the development of technology for the process of building production as the core business of construction industry. As results of technology development, many new types of buildings and structures such as large-scale space structures, seismic isolation buildings and super-high-rise buildings have been produced, and many innovative construction methods have been developed.

11.3.5 Client's satisfaction

The investigation of client satisfaction is carried out after the building project has been completed. In this investigation, the client's satisfaction is determined by a "hearing investigation" from the "owner", “orderer” and "the third person". The results of the investigation are reflected in the appropriate response to problems and to the improvement of building quality. Previously, only the project developer's satisfaction had been investigated. More recently, the satisfaction of users, neighborhoods and community have been investigated as well. These investigations are achieved by the company itself and by the third organization.

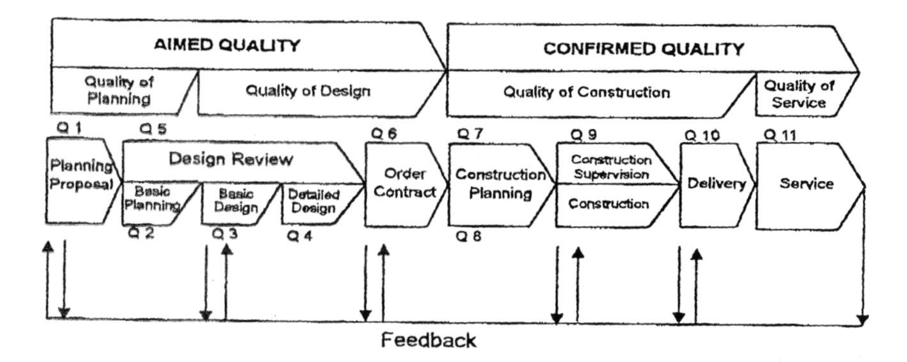

Fig.1 Flow of Building Production

11.4 Quality Management System – A Case Study for Concrete Work

11.4.1 Quality management in concrete work

The construction supervisor, constructor, specialist contractor, ready mixed concrete manufacturer and testing institute all are concerned with the quality of the concrete work. These persons and organizations are responsible for the activities for quality management and quality assurance to satisfy the requirement for the concrete work and to provide confidence to the quality of the building. The persons concerned clarify the quality management system to carry out the quality control. The role of the persons concerned with the concrete work is as follows:

Construction supervisor: The roles of the construction supervisor, who is the owner's agent, are;

1) To review design documents and specifications from the aspect of supervision,
2) To establish a supervision plan and to communicate it to the owner and constructor,
3) To review the drawings made by the constructor,
4) To verify the results of the quality management achieved by the contractor at each stage of the construction, and
5) To inspect the quality of concrete in structural body and to indicate an improvement measure to the constructor when the problem arises on the quality of concrete.

Constructor: A constructor supervises a specialist contractor and a ready-mix concrete manufacturer in placement of concrete in the building. A constructor achieves the quality control of the construction work as a part of the construction management process (a general term concerned with control or management for safety and sanitary, construction process, cost and quality) to provide a building resulting in a high degree of client satisfaction. The main activities of a constructor are;

1) To review design drawings and specifications,
2) To establish a quality control plan,
3) To investigate and select a ready-mix concrete manufacturer,
4) To order ready-mix concrete,
5) To inspect accepted ready-mix concrete,
6) To manage field construction, and
7) To inspect and to test concrete placement in the structural system.

Specialist contractor: A specialist contractor organizes all the workers concerned to satisfy the query item of the constructor and to achieve the quality management goals in the stipulated range. The result is reported to the constructor and is subjected to acceptance inspection.

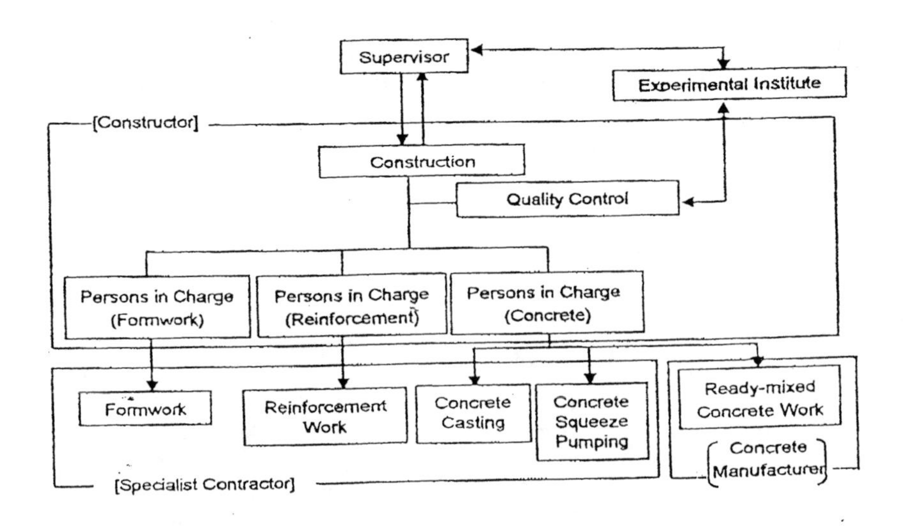

Fig. 2 Quality Management System in Reinforced Concrete Work

Ready-mix concrete manufacturer: A ready mixed concrete manufacturer clears a part assignment inside the factory to satisfy the query item of the constructor, and achieves the quality management regarding the manufacturing ready mixed concrete and the transport. Results are reported to the constructor. It is subjected to acceptance inspection.

Testing institute: The testing institute performs the tests required by the construction supervisor or the constructor and test results are reported. The constructor or construction supervisor must evaluate the testing institute from the aspect of transparency and reliability. An examination is the very critical act to get necessary data and the reliability of the data must be sufficiently high to make a decision.

11.4.2 Principles for quality management achieved by constructors

The basic goals of quality management achieved by constructors are as follows.

1) To fully grasp the required quality (design quality) of the concrete work described in the drawings and specifications and in the demands of the construction supervisor. Any unclear points are discussed with the construction supervisor, and the resolutions are indicated in the paper documentation.
2) To establish a quality control plan to cover the concrete work entirely and to submit it to the construction supervisor. The main contents of the quality management plan contain the goal of construction quality, the quality control system, the process management, the inspection and the person in charge, and so on. The verification items for the construction supervisor and the documentation and records are also included.
3) To decide upon quality control manager and to report it to the construction supervisor before the work starts. The quality control manager must possess appropriate qualifications, such as the registered architect of the first class. A quality control manager organizes the system including a specialist contractor, a ready mixed concrete manufacturer and the testing institute to attain the goal of the design quality and construction quality.
4) To achieve process management, to carry out necessary inspection and test and to report the results to the supervisor. The process of the examination, inspection and decision criterion are provided in the quality control plan document.
5) To discuss in advance with the construction supervisor the consequences of a failure in test or inspection. To promptly discuss with the construction supervisor when a failure occurs in structural concrete regarding an item not specified in the quality control plan document.
6) To examine the results of the quality control program and to document these results to reflect on the next process.
7) To maintain quality record for a necessary period. All the quality record must be kept until the construction completion. A storage period is decided as every quality record after the completion.

11.5. A Case Study of Quality Management in Construction

11.5.1 Introduction

This section summarizes the bottom-up concreting construction of a 39-story hotel building using concrete filled steel tubular (CFT) columns. The CFT columns used super-high strength concrete of the specified design strength (F_c) 60N/mm^2 and also used silica fume as cementing material. This project utilized Fc60 high strength concrete silica fume in a high-rise building in Japan for the first time. The plant equipment for powder silica fume developed to manufacture this concrete was the first example in the world.

Photo 1 Shinagawa Prince Hotel (Tokyo)

11.5.2 Building profile

This hotel building is located western Tokyo and has 1,700 guest rooms. The typical floor plan is 36m×72m, and the central portion in plan is used for banqueting halls of two-story height. The building has two basement stories, one penthouse and stands 39 stories above the ground. The total floor area is 108,535 m^2 and the eave height is 137.9m. The hotel construction occurred from July 1991 to October 1994, a period of 39 months.

In the longitudinal direction, the structure was a rigid frame, and in the transverse direction a combination of rigid frame structures with brace structures. Underground stories are steel reinforced concrete (SRC) and stories 2-5 are SRC column, steel beam structures. The stories above 4th floor are steel structures and concrete filled tubular (CFT) columns were adopted in the interior frames of the 4th to 20th floor. The floor slabs and exterior walls are of precast concrete.

In the preliminary design, the weight of the structural steel in the columns exceeded 50% of the total weight of steel; therefore, the steel columns in the 4th to 20th stories, which were subjected to high axial load, were replaced with CFT columns to reduce steel volume. The super-high strength concrete was adopted to increase the axial load carrying capacity of concrete and, as a result, to reduce steel volume of CFT columns subjected to very high axial loads. Steel volume was significantly decreased by the adoption of super-high strength concrete. The contribution of the decrease in steel volume to the total construction cost was appreciable even if the increase of design stress and the addition of concrete construction due to the increase of the building weight were counted.

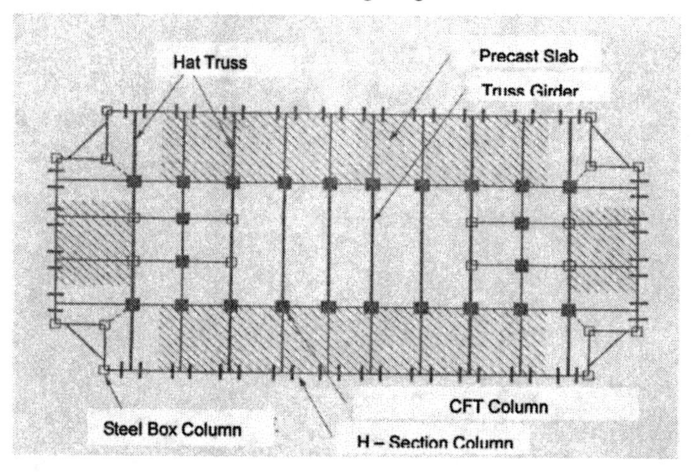

Fig. 3 Structural Plan of Typical Story

11.5.3 Silica fume concrete and manufacturing

Silica fume reacts with Ca $(OH)_2$ produced by the hydration of cement to produce calcium silicate hydrates. The grain diameter of the silica fume is about 1/100 of the diameter of cement and its effect on filling the space between cement grains is high. The silica fume increases concrete strength due to these physical and chemical actions. Moreover, the L-flow speed of super-high strength concrete increases and viscosity decreases by the addition of silica fume. Thus, very workable super high strength concrete in concrete filling, concrete casting and squeeze pumping can be realized using silica fume.

The plant equipment for powder silica fume (SFC plant) was developed and provided for the construction of this building following four basic principles:

1) It is prepared in an existing concrete-mixing plant.
2) It works with the control plank of the batcher plant together, and the measure and stripping of silica fume are done with the control chamber.
3) It has the measure precision specified in the Japanese Industrial Standards.
4) Its removal, assembling and conveyance are easy, and it can be provided in any ready-mix concrete plant in Japan.

When silica fume is stored in an ordinary cement silo, it flow under gravity when the gate is opened. Therefore, the developed silo adopted the mechanism to discharge silica fume by a screw feeder and the silo was set in horizontal form. An air squeeze pumping by high air pressure was adopted. Silica fume was mixed with air in the blow tank and was sent with air pressure to the meter mounted in the batcher plant. A meter was mounted at the top of mixing machine and the silica fume measured by the air slide-type discharge mechanics was discharged into the mixing machine. An inaccuracy measure prevention device was mounted in the meter.

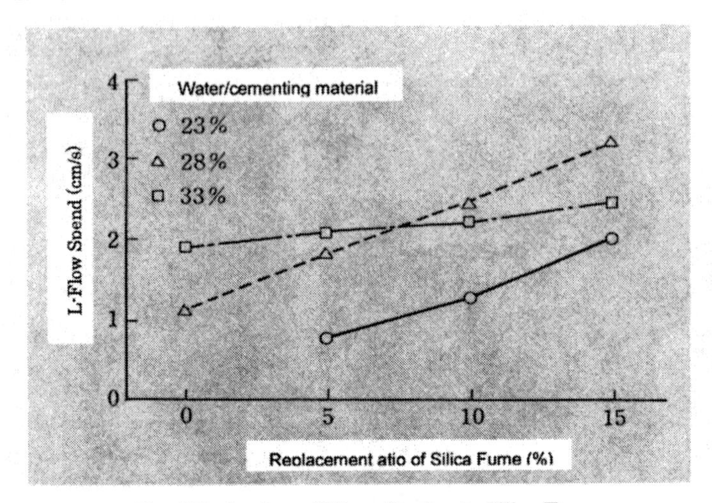

Fig. 4 Reduction of Viscosity due to Silica Fume

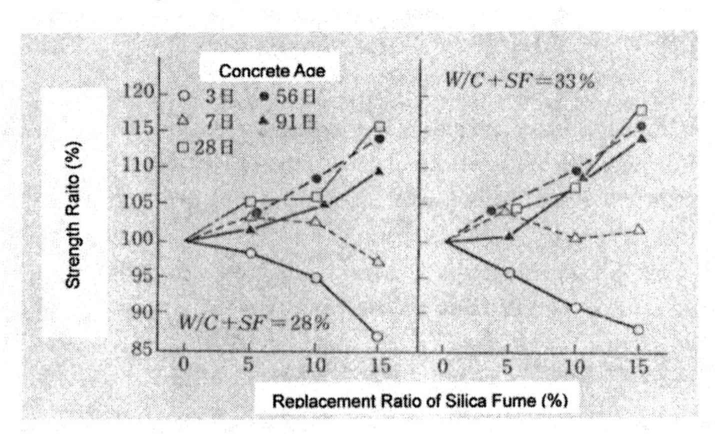

Fig. 5 Enhancement of Strength due to Silica Fume

The control plank of the SFC plant was mounted in the control chamber of the fresh concrete mixing plant, and worked with the control plank of the batcher plant together, to automatically measure and discharge silica fume. The specifications of SFC plant are as follows:

The capacity of storage: silica fume 4 tons (to be used for concrete volume $70m^3$)

Maximum measure value: 120kg
Measure precision: ± 2%,
Cycle time: 60 second/cycle

Photo 2 Silica Fume

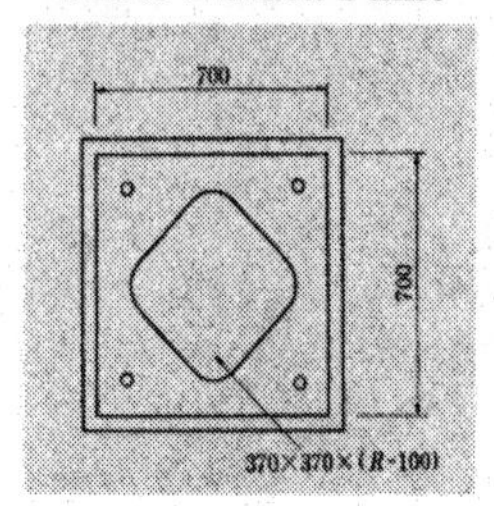

Fig.6 Diaphragm of CFT Column

11.5.4 Bottom-up concreting construction

The opening of the diaphragm in a steel tube column used for bottom-up concreting was square in its shape and rotated at 45° with respect to the column profile. The opening was 26% of the column section area. As large an opening as possible was provided so that the resistance to bottom-up concreting might be made small. Four round openings of diameter 25mm were set up approaching at the corners of the diaphragm. This was the air vent for the bubble formed between the diaphragm bottom surface and the concrete top surface at the time of bottom-up concreting. It was necessary to understand the difference in the strengths of standard curing specimen and the concrete in structures because the concrete compaction was not done. And any void might not be formed on the bottom edge of the diaphragm.

It was necessary to evaluate the bottom-up concreting load and the load of squeeze pumping between pumping car and the mouth for bottom-up concreting (increase of pressure to the liquid pressure) corresponding to the diaphragm shape. This field construction of the Fc60 super-high strength concrete, even using silica fume, was the first such experience in Japan. Therefore, the plans of concrete mixing, squeeze pumping and bottom-up concreting were established after a series of preliminary laboratory tests of concrete mixing, concrete strength in structural frame and the workability in construction described below.

Squeeze pumping test: A squeeze pumping test was done for the silica fume concrete with cementing material ratio 23-33% and slump 23cm to measure squeeze pumping load. The replacement ratio of silica fume was 10%. The material used was the same as that for actual construction. The conveying pipe was 127 mm in diameter and 165m long. The used pump was cylinder type and the maximum discharge capacity was $60m^3/h$ and the estimated discharge pressure was $60kgf/cm^2$. The cylinder was $\varphi 205\times 1650mm$

The pressure loss varied with the L-flow speed (concrete viscosity) for the concrete of 23cm slump under the same discharge magnitude. As a result of the test, it was recognized that the viscosity in terms of the L-flow speed should be considered as well in case of super high strength concrete to calculate squeeze pumping load though the pressure loss has been evaluated only by the slump and the discharge magnitude.

Bottom-up concreting construction test: A bottom-up concreting construction test was carried out to measure the bottom-up concreting load, concrete filling appearance on the bottom edge of the diaphragm and the structure body strength using the steel-pipe column test body of 9.85m long with the same section to be used for actual construction. The concrete with the slump flow 45cm, the water cementing material ratio 28% and the slump 23cm was used. The 10% silica fume to cementing material was added. The bottom-up concreting speed was $20m^3/h$.

Concrete quality gradation due to the bottom-up concreting was small, although some decreases were recognized in the flow value and the L-flow speed. The pressure in the piston stop of the pump car almost corresponded to the liquid pressure. The difference in the maximum pressure in the piston operation and the liquid pressure corresponded to the bottom-up concreting load. Total pressure in the bottom-up concreting was 1.13 times of the liquid pressure.

Settlement of concrete 3-4 hours following the bottom-up concreting was observed to be 1.3mm. The observed core strength was 63-76 N/mm^2 and the tendency that the bottom becomes stronger was recognized. The difference between mean value of the core strength and the strength of standard curing specimen was $12.7N/mm^2$. A test body was dismantled after core drilling and the filling appearance of the bottom surface of diaphragm was observed. Though a small bubble of about 3% of the total area was observed, diaphragm bottom surface and the concrete adhered well and large voids and settlement were not observed.

Mix proportion planning: The concrete mix proportion was determined based on the following formula, referring to the guidelines proposed from the "New RC" research project, so that the structural concrete strength would meet the specified design strength at the age of 28 days.

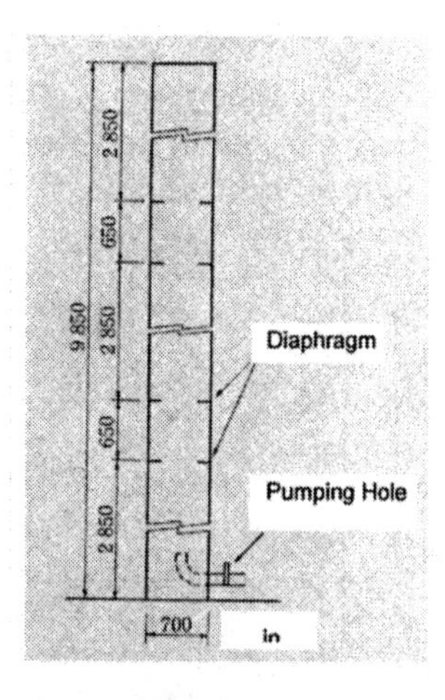

Fig.7 Steel Tube Specimen for Squeeze Pumping

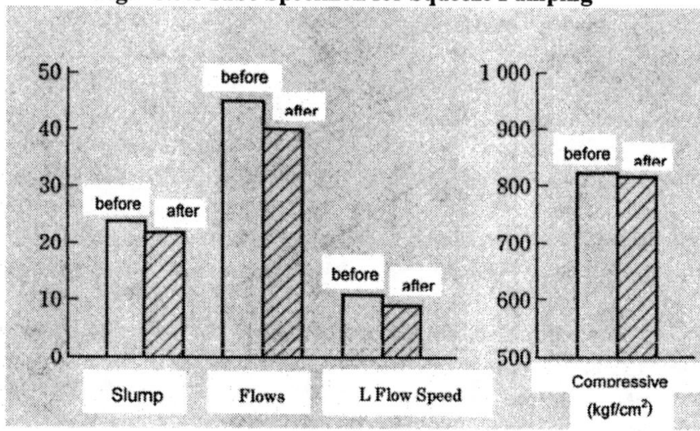

Fig. 8 Change of Concrete Quality after Pumping

$$F_{28}=Fc+S+K\sigma \quad (1)$$

Here, F_{28}: proportioning strength, Fc: specified design strength, S: difference of the strengths in standard curing specimen and structural body, σ: standard deviation of concrete strength, K: normal deviation. The factor S was set as 14.5N/mm^2 referring to the construction test above. Standard deviation was set as 8% to Fc plus S. Normal deviation K was determined based on the defective ratio specified in the JASS5 (Japanese Architectural Standard Specification, Reinforced Concrete Work), and the proportioning strength F_{28} was determined as

$86.5N/mm^2$ from these conditions.
The concrete mix proportion was determined by experimentation to provide as much fluidity as possible. The slump and the slump flow were 23 cm and 45cm, respectively. To decrease the viscosity, silica fume of 10% to connection material was added. Pit sand (specific gravity 2.61, coefficient of water absorption 1.53% and FM2.99) was as a fine aggregate and the crushed limestone (specific gravity 2.70, coefficient of water absorption 0.38% and FM6.42) was used as coarse aggregate. The high performance AE water reducing agent with special sulfone radical carboxyl group containing multiple dimension polymer was used to reduce volumetric change of concrete in steel tube.

Squeeze pumping and bottom-up concreting plan: The columns of 61.9m from the level of the 4th floor to the 21st floor were the CFT columns. A concrete pump car was mounted in the second floor. The 127mm pipes were arranged from the 2nd floor to the bottom-up concreting mouth at the 4th floors. It was planned to achieve the bottom-up concreting through the height of 61.9m at one time. Horizontal piping was maximum 57m and vertical piping was maximum 9.8m.

Results of bottom-up concreting at site: The test of fresh concrete was carried out once in every $30m^3$ at shipping and in $60m^3$ at unloading at site. The variances of the quality of fresh concrete and compressive strength were small. The mean of the compressive strength was $3.5\text{-}4.5N/mm^2$ higher than the proportioning strength. The standard deviation of the compressive strength was $4.2 - 4.4N/mm^2$ (coefficient of variation 4.9%) in terms of the standard deviation and it was smaller than the design value $60kgf/cm^2$ (8% of FC+S) established at the time of the mix plan. Though concrete conveyance time was about 25 minutes, quality degradation at unloading was not detected. Though the bottom-up concreting of 61.9m took 1 hour the concrete held good fluidity after the bottom-up concreting as well. The gradation of the compressive strength was not recognized, either. The settlement at the top of concrete finished in 5 hours after the bottom-up concreting, and it was very small value with 1.5mm. This was about the same value as that in construction test described above. The axial strain of steel tube column to the whole length corresponding to this settlement was 20×10^{-6} and 500×10^{-6} when the strain is concentrated in one-story column length. Thus the settlement of concrete was very small.

Photo 3 Appearance of Concret Beneath Diaphragm

Photo 4 Squeeze Pumpin

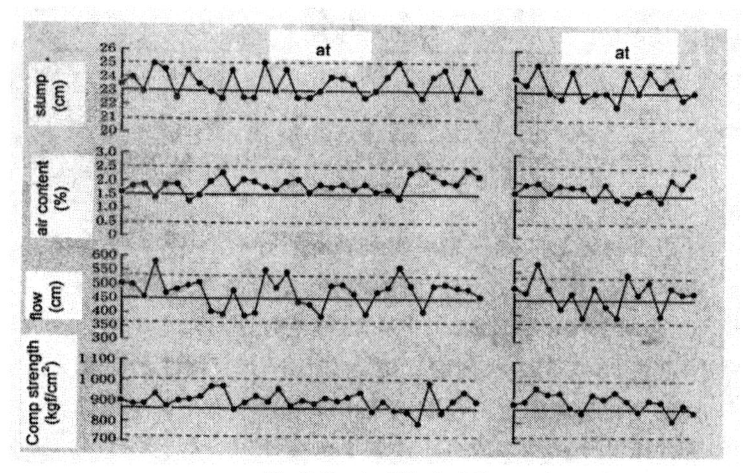

Fig.9 Control Test of Concrete

11.5.5 Concluding remarks

The construction process achieved stable manufacturing of super high strength Fc 60 silica fume concrete and smooth bottom-up concreting. The bottom–up concreting to the height of 61.9m was achieved as planned. The technology development and preliminary work, such as the development of silica fume concrete plant, various tests for concrete mix proportioning and construction, needed four years to implement. The result of this construction was not limited to concretes with a specified design strength 60N/mm^2. It may be extended to the super-high strength concrete of specified design strength 100N/mm^2. The super-high strength silica fume concrete was put to practical use in this construction project. Moreover, high strength concrete has significant potential as an innovative structural material in new large-scale structures such as super-high-rise buildings, large-space structures and long span bridges.

References

Architectural Institute of Japan (1993),"Japanese Architectural Standard Specification JASS 5 Reinforced Concrete Work, 1993 Version," (in Japanese)

Architectural Institute of Japan (1999),"Recommendation for Quality Control in Concrete Work," (in Japanese)

Hiroyuki Aoyama (1999),"Development of High-rise Concrete Construction in Seismic Countries," George Winter Commemorative Lecture, 1999 ACI Fall Convention

Japanese Standard Association (1993),"Japanese Industrial Standards – JIS A 5308:1993 Ready-mixed Concrete," (in Japanese)

Japanese Standard Association (1994),"International Standards for Quality Assurance – A Full Translation of ISO Standards and Commentary," (in Japanese)

Toshio Yonezawa, et al (1993)," Bottom-up Concreting into Steel Tube Columns Filed with Ultra-High-Strength Concrete Using Silica Fume," Concrete Journal, Vol.31, No. 12 (in Japanese)

Ryuzou Nakagome (1997),"Building Production and Quality Management in Takenaka Corporation," A critical factor is "quality" and persistent challenge, Standardization and Quality Control, Vol.50, No.7 (in Japanese)

Chapter 12

Human Error in Constructed Projects

By:
Karen C. Chou, Minnesota State University, Mankato

Abstract

This chapter presents an overview of the types of human error occurring in civil construction and how these errors could be minimized. Four failure cases are presented to illustrate human error induced failures and the lessons learned from them. The goal of this chapter is to improve the quality control and assurance of engineering design and construction through some of the lessons learned from the sample failure cases. A list of references is included for readers who wish to explore more about human error in constructed projects.

12.1 Introduction

The first Canon of the Code of Ethics adopted by the American Society of Civil Engineers (ASCE) states that "Engineers shall hold paramount the safety, health, and welfare of the public and shall strive to comply with the principles of sustainable development in the performance of their professional duties". This Canon clearly defines the civil engineers' role in civil construction. To assist civil engineers in fulfilling this obligation, design specifications have been written to include uncertainties in loads and resistance, and modeling of engineering behavior. These specifications, however, are the minimum requirements for the design of a safe structure. They do not include the effects of human error that may occur in planning, design, construction, and utilization of the structure (Matousek, 1981; Ellingwood, 1987).

Failures rarely occurred due to the stochastic nature of loads and resistances, or the formulation of engineering mechanism (Ellingwood, 1987). If a structure was used for a function differed from its original intent and the structure failed because the loads were significantly larger than the initial design loads, then the failure was due to "an error in utilization". If the failure mode was not the one that was considered, or not fully understood during the design process, then the failure was due to "an error in design". Errors of this nature are not reflected in the factor of safety commonly applied in design specifications or design practice. These errors can be classified as human error.

Although the loss of lives in a structure failure can be high (114 deaths resulted from the Hyatt Regency walkway failure in 1981 (Pfrang and Marshall, 1982)), throughout history, the occurrence of structure failure is rare considering the amount of exposure we have with structures. Hence, the risk of death from structure failure is relatively small especially when one compares it with the risk of death of other societal activities (Melcher, 1995). For example, the risk of death per year from alpine climbing is 15,000 to 20,000 times as high as that from structure failure; it is 10,000 times as high from cigarette smoking; and 2,000 times as high from car travel. However, the public's reception of casualties of structure failure is quite different from that of a car accident or cigarette smoking. The expectation from the public is that the structures (buildings and bridges) have been around for a long time and should remain for a long time to come. The public does not expect these structures to fail (Melcher, 1995). Whenever a structure fails, especially when human lives are involved, the public's confidence in structures and the engineering behind them is shaken. Hence, as rare as structure failures have been and as low risk of death as studies have shown the civil engineering profession is still seeking ways to reduce these failures, especially human error induced failures.

In this chapter, four well known failure cases occurred in the United States between 1976 and 1986 are examined. Lessons learned from these failures and there relationships to earlier chapter (chapter 10) are discussed. A brief discussion of the research conducted in the modeling and reliability analysis of human error is also presented. A list of bibliography is included in the Appendix for readers who are interested in other structure failures as well as studies done to assess or model them.

12.2 Sources of Human Error

Nowak (1992) defined human error as a departure from acceptable practice. Melchers (1995) considered human error as factors that lead to a structure failure and these factors are a direct result of human input or lack of it. Ignorance, negligence, carelessness, and lack of knowledge are samples of human error.

Blockley (1977) identified eight basic causes of failures of ordinary construction where an element of human error existed. They are: (1) inadequate consideration of uncertainty in design variables; (2) errors in methods of structural analysis; (3) extreme hazards such as flood or earthquake not considered; (4) failure modes not understood; (5) failure modes recognized but not treated properly; (6) mis-management or poor communication during construction; (7) financial or political pressures on personnel; and (8) misuse or wilful abuse.

Nowak and Carr (1985) grouped the basic error types to three: (1) errors of concept; (2) errors of execution; and (3) errors of intention. The first type includes loads not envisioned, incorrect assumptions, or incorrect use of analytical modeling. The second type includes calculation or detailing errors, mistakes in reading drawings and specifications, and defective workmanship. The third type includes unwarranted shortcuts to save money, substitution of components that may not give equivalent performance as originally specified, and acceptance of marginal workmanship due to construction schedule (Ellingwood, 1987).

Although the above causes of failures were based on studies of ordinary constructions, the causes can occur in special structures such as tall and super-tall buildings. The consequence of these errors on those special structures will likely be amplified. Furthermore, there exist additional sources of failures unique to the design and construction of special structures, for example, use of new materials or design concept that has not been used in other structures.

12.3 Case Studies

The best thing that comes out of any failure is the experience or lessons learned to prevent that types of error from recurring in the future. Four well known failure cases occurred between 1976 and 1986 are presented herein to illustrate the consequence of human error and the lessons learned. These cases are: (1) the Hyatt Regency Hotel walkway collapse in 1981; (2) the Hartford Coliseum roof collapse in 1978; (3) the Teton Dam failure in 1976; and (4) the space shuttle *Challenger* accident in 1986. Although the *Challenger* explosion was not a civil structure failure, it presented human errors that can be found in special structures. Lessons learned from these failures may help engineers prevent future structure failures.

In each of the cases presented, the failure was briefly described. Detail description of the failures can be found in the references cited for each case. The findings from the investigative team(s) are presented. The lessons learned or recommendations for improvement are presented at the end of each case.

12.3.1 *Hyatt Regency Hotel Walkway Collapse in Kansas City, Missouri - July, 1981 (Ross, 1984 and Pfrang and Marshall, 1982)*

On July 17, 1981, two indoor walkways of the Hyatt Regency Hotel in Kansas City collapsed killing 114 people and injured 185 (Pfrang and Marshall, 1982). At the time, it was the most devastating structural collapse ever to take place in the United States.

The atrium lobby of the Hyatt Regency Hotel consisted of three walkways that spanned about 120 ft at the second, third, and fourth levels. The fourth level walkway was directly over the second and the third-level walkway was offset to one side. The third and Fourth level walkways were each held up by six 1-1/4-in diameter steel rods that were suspended from the ceiling and connected to transverse box beams. These box beams were made up of pairs of MC8x8.5 channels butted toe to toe and welded longitudinally. The longitudinal stringers were W16x26 and were bolted to the clip angles which were welded to the box beams. A steel deck with 3-

1/4-in thick light weight concrete slab acted compositely with the W16x26 beams. The ends of each box beam at the fourth level had two holes drilled to allow the hanger rod from the ceiling to pass through one hole and another hanger rod to suspend from the second hole to connect to the second level walkway.

A comprehensive laboratory testing program was conducted at the National Bureau of Standard NBS (now called the National Institute of Standards and Testings) (Pfrang and Marshall, 1982). The test results indicated the action of loads at the time of walkway collapse was substantially less than the design loads specified by the Kansas City Building Code. The test concluded that the quality of workmanship and the materials used in the walkway system did not play a significant role in initiating the collapse. Investigators at NBS concluded that two factors contributed to the collapse: inadequacy of the original design for the beam-hanger rod connection and a change in hanger rod arrangement during construction.

In the original contract drawings, the hanger rod connection called for the 1-1/4-in diameter rod traveling from the ceiling through the box beam in the fourth level and connected to the second level box beam. A nut and washer were used underneath each box beam to tighten the hanger rod connection. According to published reports, it was unclear where the change was initiated and by whom. All parties involved had either denied the approval of the change or declined any comments. However, based on the shop drawings published by the *Kansas City Star,* the shop drawings showed the stamps and initials from the hotel's architect, the structural engineer of record, and the contractor. This still did not prove where the change was first proposed. It did indicate the awareness by the three parties on the change.

The tests conducted at NBS concluded that the change in the hanger rod arrangement had essentially doubled the load on the box beam-hanger rod connection at the fourth level walkway. Moreover, the original connection design (use of a continuous rod shown in the contract drawing) would not have satisfied the Kansas City Building Code. However, it would have supported the actions of loads at the time of collapse.

The Kansas City Hyatt Regency Hotel failure was perhaps one of the well publicized failures at the time. The incident led, at least indirectly, to the development of a guideline on constructed projects by the American Society of Civil Engineers (ASCE, 1988). Nearly 20 years after the failure, engineers including the engineer of record of the project, re-examined that fateful incidence back in July 1981. The lessons learned from the failure investigation and suggestions offered were: (1) Communications among all parties of the design team is imperative (Gillum, 2000). The study of the design process indicated a failure in communication at the transfer of responsibilities for the designs of specific elements of the structure from one design/detailing team to the next (Luth, 2000). (2) The lack of redundancy in overall design of the walkway support system. Failure of all connectors was imminent. Failure of any one connector would lead to progressive failure and total collapse. There was no reserve capacity at the connections (Pfrang and Marshall, 1982). (3) The responsibilities of each party of the design team should be defined and known to all parties involved. It was suggested a nationwide standard of practice should be adopted (Gillum, 2000; Moncary and Taylor, 2000). ASCE has attempted to do that with the publication of the *Quality for Constructed Projects: a Guideline for owners, designers, and contractors* (ASCE, 1988).

12.3.2 Hartford Coliseum in Hartford, Connecticut – January 18, 1978 (Smith and Epstein, 1980; Ross, 1984)

The roof of the Harford Coliseum collapsed at approximately 4 am on January 18, 1978. Fortunately, there was no one in the building at that time. No deaths or injuries were reported. If the collapse were to have occurred six hours earlier, the casualties would have been tremendous and there were thousands of spectators watching sporting event just hours prior to the collapse.

The roof of the Hartford Coliseum was a space truss covering an area of 360 ft by 300 ft. The clear span was 270 ft by 210 ft. The roof was supported on four pylons near the corners. The truss was 21 ft deep. Top and bottom chords were arranged in a 30 ft by 30 ft grid. Main diagonals were also 30 ft long. Intermediate bracing 15 ft in length was connected at midpoints of the main members. Most of the main members were composed of four equal-leg steel angles shop-bolted back to back to form a built-up cruciform shape. The angles were separated by spacer plates for end connections at the gusset plates. There were as many as eight members connected at an end connection, each with three components of slope. The truss was flat. The dome-like rooftop was achieved by supporting a roof skin on stub columns above the top chords of the truss. The 2300 member roof structure was thought to be highly redundant structurally.

The construction of the Coliseum was completed in 1973 (Feld and Carper, 1997). It was built in a nontraditional fast-track manner with the contract subdivided into five small contracts, all coordinated by a construction manager. The steel space truss assembly and erection was one of the five contracts.

At the time of the collapse, the Coliseum had experienced the heaviest snow fall in its five years of service. The roof was designed for a snow load of 30 psf. The investigation by the U.S. Army Cold Region Research and Engineering Laboratory performed on the day after the collapse across the street from the (Coliseum was words are missing here) weighed 23 psf (Ross, 1984).

Numerous investigation teams were assembled by the various parties involved in the project. The Hartford City Council appointed a 3-member investigation panel who in turn hired Lev Zetlin Associate of New York City (LZA) to perform the investigation. The mayor of Hartford commissioned an academic task force to conduct an independent investigation. The findings reported herein are primarily based on the investigations done by the academic task force and LZA.

Both investigations, although differing in their models and analyses, agreed that the lack of or insufficient support to lateral stability in some of the compression members of the space truss played an important role in the roof failure. In traditional roof structure, the roof skin would have provided some lateral support to the space trusses. In the Hartford Coliseum, with the roof skin set above the space truss, the lateral supports of the space truss had to be provided specifically (Kaminetzky, 1991). The LZA report indicated some compression members had an unbraced length of 30 ft instead of 15 ft assumed in the computer analysis. There was also some unanticipated eccentricity found at the complex connections. The sudden collapse also raised questions on the roof truss's vulnerability to progressive collapse and its redundancy (Smith and Epstein, 1980).

During the failure investigations, it was found that there were excessive deflections, a sign of structural deficiencies, observed during the construction process. The measured deflections, some were as much as twice the value predicted by the computer analysis could have been a warning that a problem existed. In the report to the Hartford City Council, LZA stated that "there should have been a resident specialist structural engineer with full-time involvement during the assembly and erection of the space truss roof." (Ross, 1984). During the investigation, the Hartford City Council's panel found that the architect had twice recommended that a qualified structural engineer be hired to provide on site inspection (Feld and Carper, 1997). The recommendation was rejected by the construction manager.

Among the lessons learned from the investigation of the roof collapse of the Hartford Coliseum was: (1) the responsibility of the architects, engineers, and construction manager should be well defined. Are the architects and engineers responsible for performing traditional services, such as field inspection, even when their contracts and fee agreement are based on reduced service (Ross, 1984; Feld and Carper, 1997)? (2) A computer is simply an analytical tool (Feld and Carper, 1997). All design assumption must be diligently checked with as-built conditions, and actual measured field performance should be compared with predictions. The bottom chord deflection during the erection of the truss in January 1973 was recorded at 8.4 inches under dead load alone. The measured deflection had exceeded the total design load deflection of 7.35 inches despite the fact that other dead loads such as permanent roofing and cementitious deck were not in place. By April 1975, the measured deflection was 12 to 13 inches. The design calculation predicted that it should have been 8.5 inches. (3) The space truss was considered to be highly redundant. Should the collapse have been as sudden as observed? How well were the loads being redistributed when one member fails (Smith and Epstein, 1980)? (4) Is the use of traditional design factors of safety adequate for projects built using nontraditional approaches to project delivery, and for structures such as the sports complex where a lot of people can be assembled at one time (Kaminetzky, 1991; Feld and Carper, 1997) ?

12.3.3Teton Dam, Idaho – June 5, 1976 (Ross, 1984

The collapse of Teton Dam in Idaho on June 5, 1976 marked the first catastrophic failure ever experienced by the U.S. Bureau of Reclamation (BuRec) in over 75 years of successful dam construction (Feld and Carper, 1997). The first sign of trouble was observed on June 3, 1976 on the right bank downstream from the spillway with a see page of 20 gpm (gallons per minute). The rate had not changed until the fateful day of June 5. The seepage rate went from 20 gpm to 50 to 60 cfs (cubic ft per second) at 9 am. At 9:30 am the rate was estimated at 70 cfs. Evacuation was ordered at the towns downstream of the dam. At 10:30 am, the seepage through the embankment had reached 1000 cfs and a whirlpool in the reservoir indicated failure was imminent. Somewhere between 11:30 am and noon, the embankment failed and released a wall of water 15 ft high. The failure of Teton Dam resulted in the loss of 11 lives and millions of dollars in property damage (BuRec, 2002).

Teton Dam was a 305 ft high earth-filled dam with a 3050-ft long embankment comprised of 10 million cubic yards of earth and rock fill. The dam was founded on a basalt formation containing rhyolite, a porous rock with considerable fracture and voids. An extensive grouting program was specified by the Bureau of Reclamation.

A curtain of concrete 3000 ft long and as much as 300 ft deep was constructed beneath the trenches and extended 1000 ft beyond the abutments. Keyways 70 ft deep and 30 ft wide at the bottom were excavated into relative stable rock in both abutments. A core trench across the canyon floor linked the keyways. The grout curtain was injected through three rows of holes, upstream and downstream along the core trench on 20-ft centers, and through a central concrete grout cap on 10-ft centers. On the canyon bottom, directly under the former riverbed, all grout holes were 10-ft center-to-center. The core trench in this area was excavated 100 ft to sound rock.

Two teams, a 10-member external independent panel and the internal Interior Review Group (IRG), were charged to conduct the investigations. Both teams concluded that the principal cause of the Teton Dam failure was deficient embankment design and inadequate field inspection by the Bureau of Reclamation. The external panel completely exonerated the prime contractors. In an updated analysis by the Interior Review Group in 1977 and 1978, IRG found instances where the contractors did not follow the specifications. However, the infractions alone were not significant enough to cause the failure.

Since the failed portion of the dam was washed away, the cause of the failure may never be established with certainty. The external panel cited that the BuRec's design proceeded "without sufficient consideration of the effects of differing and unusually difficult geological conditions at the Teton dam site" (Ross, 1984), and that the dam was not sufficiently instrumented to warn of trouble. The panel concluded that the causes of failure were: (1) Specific measures were not taken to insure sealing of the upper part of the rock under the grout cap. Water was able to move through the highly jointed volcanic rock. (2) The impervious material (zone 1) in the core of the dam was not adequately protected by the layer of semi-pervious material (zone 2) to provide drainage. The tests showed that the zone 2 material was almost as impervious as zone 1 material. (3) The designers emphasized keeping the water from seeping through the dam rather than taking measures to "render harmless whatever water did pass".

The final report by the Interior Review Group supported the findings of the external panel. The design deficiencies presented in the Teton Dam were the results of engineering judgment of what design process and inspection procedures should be followed. The external panel found that the site selection process and geological studies of the dam site were appropriate and extensive.

The Teton Dam failure has led the Bureau of Reclamation to modify its policy on design process and operations of dams. The modifications included: (1) to commission independent consultants to review the design of all major new dams. (2) To increase instrumentation during construction. (3) To consider the need for greater outlet capacities to provide better control over reservoirs.

12.3.4 Space Shuttle Challenger Accident – January28, 1986 (Rogers Commission, 1986)

About 72 seconds after liftoff, the space shuttle *Challenger* exploded. All seven crew members, including the first and only civilian mission specialist, a school teacher, perished.

According to video recording of the launch, the first sign of trouble was observed at 0.678 seconds into the flight. A strong puff of gray smoke was spurting from the vicinity of the aft field joint on the right Solid Rocket Booster (SRB). The first flickering flame appeared on the right SRB, again, in the area of the aft field joint was detected at 58.788 seconds into the flight. It grew into a continuous, well-defined plume at 59.262 seconds. At about the same time (60 seconds), telemetry showed a pressure differential between the chamber pressure in the right and left boosters. The right booster chamber pressure was lower, confirming the growing leak in the area of field joint. Beginning at about 72 seconds, a series of events occurred extremely rapidly that terminated the flight.

Under the Executive Order of the President of February 3, 1986, a Presidential Commission was appointed to investigate the space shuttle *Challenger* accident. The findings presented here were extracted from the Commission's report (Rogers Commission Report, June 6, 1986). A portion of the report is available online (see reference section for details). The Commission's findings and recommendations were extensive including issues not directly related to the accident, but which may affect the safety of future shuttle missions. The findings presented here only pertain to the cause of the accident.

The consensus of the Commission and participating investigative agencies is that the loss of space shuttle *Challenger* was caused by a failure of the pressure seal in the aft field joint (joint between two lower segments) of the right Solid Rocket Motor (also referred to as the Solid Rocket Booster). The specific failure was the destruction of the seals that were intended to prevent hot gases from leaking through the joint during the propellant burn of rocket motor. The failure of the seal was due to a faulty design unacceptably sensitive to a number of factors. These factors were the effects of temperature, physical dimensions, and the character of materials, the effects of reusability, processing and the reaction of the joint to dynamic loading.

Moreover, the Commission concluded that there was a serious flaw in the decision making process that contributed to the accident. Those who made the decision to launch the *Challenger* were unaware of the history of problems concerning the seals (O-ring) and the joint and were unaware of the initial written recommendation of the contractor of the Solid Rocket Booster advising against the launch at temperature below 53°F (the ambient temperature at launch time was 36°F, 15 degrees lower than the next coldest previous launch) and the continuing opposition of engineers of SRB after the management reversed its position.

The Commission also found lack of involvement by the safety, reliability and quality assurance unit within NASA. The Commission cited: (1) the safety program was understaffed, and posed a conflict of interest as the unit was under the supervision of the organizations and activities that the safety unit has to check. (2) No trend analysis was performed on problems experienced by the O-ring from prior launches. (3) Problems reporting requirements were not concise and failed to get critical information to the proper level of management.

The recommendations presented by the Commission were extensive. They included design, shuttle management structure, critically review and hazard analysis, communications, landing safety, launch abort and crew escape, flight rate, and maintenance safeguards. Pertaining to the cause of the *Challenger* accident, the Commission's recommendations were: (1) the faulty Solid Rocket Motor joint seal must be changed. No design options should be prematurely precluded because of schedule, cost or reliance on existing hardware. An independent oversight committee should be formed to oversee the design effort. (2) A redefinition of the Shuttle Program Manager's responsibilities is essential. This redefinition should give the Program Manager the requisite authority for all ongoing Space Transportation System, where the Shuttle Program is part of, operations. At the time of the Challenger accident, the project managers for various elements of the shuttle program felt more accountable to the space center management than to the shuttle program organization as vital program information frequently by pass the shuttle program management.

12.4Suggestions for Minimizing Human Errors

From the four failure cases presented, three factors stand out: (1) communication; (2) responsibilities; and (3) design. In the following paragraphs, these three factors are discussed in more details. References are made between the case studies and quality assurance management presented in Chapter 10.

12.4.1 Communication

When multiple teams are involved, whether it is a complex project like the space shuttle program or the design and construction of an ordinary building or bridge, interface between different teams having input into the design process needs to be defined so that the necessary information flows across them and also so that the division of design responsibility is clear (Booth, Chapter 10). The space shuttle *Challenger* accident may be avoided if a better communication structure existed according to the recommendation of the Presidential Commission (Rogers Commission, 1986) investigating the accident. It was evidenced that the test results and conditions of the O-ring and the aft field joints observed after each launch were not communicated to the proper level of management so that an informed decision can be made.

In the case of the Hartford Coliseum, better communication between the construction manager and the design team may have discovered the design errors during the construction. In the case of Hyatt walkway, communication was believed to be one of the causes of failure (Luth, 2000; Gillum, 2000). However, it is not clear if communication was a major factor in this case. Evidence presented to the administrative hearing commission of Missouri regarding the professional conduct of the two engineers involved in the design of Hyatt Regency Hotel indicated a different twist. Evidence indicated the architect, the detailer, and the engineers' own technician did bring the change of connection design to the attention of the engineer. It was the inaction of the engineer that played a big role in the cause of failure (Rubin and Banick, 1987).

12.4.2 Responsibilities

Whose responsibility was it to design the connections of the Hyatt Regency Hotel walkway? Whose responsibility was it to interpret or to report to the appropriate team the large deflection of the Hartford Coliseum roof trusses observed during construction? These were the debates during the failure investigations. In the United States, and perhaps some other countries, the matter of *responsibility,* when failure occurred and specially when human lives are involved, becomes a legal issue. The decision is usually made by the judges or juries.

In both the Hyatt walkway and the Hartford Coliseum cases, one of the clear lessons was that a well defined responsibility of each party involved in the design and construction process must be known to all parties involved. Even in the case of *Challenger*, the Presidential Commission recommended that the shuttle program manager's responsibility should be well defined. Booth also suggested in Chapter 10 that the division of design responsibility between different groups in the design process must be clear.

Currently in the U.S., there is no national adopted standard of practice. ASCE has attempted to do that with the publication of the *Quality for Constructed Projects: a Guideline for owners, designers, and contractors*. The Guideline was first published in 1988. The second edition was published in 2000 (ASCE, 2000).

12.4.3 Design

Engineers have to make decisions or apply judgment in the use of engineering models for the analysis and design of the structure. For those projects that have precedent, the selection of an engineering model would be less challenging. For special structures, such as super-tall building, super-long bridges, off-shore structures, nuclear structures, and the space structures, there is little or no precedent. To select an appropriate engineering model becomes more challenging. Oftentimes, assumptions have to be made or scale tests have to be performed (e.g. wind tunnel test to studies the wind effects on tall buildings or long bridges). As new materials and new technology are developed, or the structural behaviors are better understood, extra care is needed when applying the new knowledge to a project. Furthermore, as Booth pointed out in Chapter 10, engineers should ensure that the structure has adequate robustness and redundancy.

In both the Hyatt walkway and Hartford Coliseum cases, the lack of redundancy gave no warning to the patrons in the hotel as the walkways crashed down to the atrium. Similar sudden failure occurred at the Hartford Coliseum.

All four projects presented here involved elements of "innovation" or unprecedented design concept or design condition. Should more attention be given to the details of the hanger-rod connection in the case of Hyatt walkways? A suspended walkway was a novel idea in the 1970s. Should the engineers be more involved in the construction process when a complex connection design was used in the Hartford Coliseum roof truss? At the time, the Coliseum was one of the biggest space frame structures in the U.S. In the Teton Dam project, the engineers faced a dam site that had considerable fractures and voids. Although the geological studies were appropriate and extensive (Ross, 1984), the Bureau of Reclamation who designed the dam did not take the necessary precautions to monitor the dam with unusual foundation conditions. The space shuttle program pushes technology to a new frontier. Extensive testing of material properties and synthesis of engineering concepts should be conducted prior to actual application. What the space shuttle program faced in applying technology and engineering knowledge to a new frontier can occur in civil constructions, such as the next tallest building, the next longest bridge, the next largest dam, or the next largest arena, where a new frontier in structural design and construction is also being explored.

Suggestions have been made to help minimize the errors involving design. They include: (1) self checking; (2) overview checking; and (3) independent checking (Booth in Chapter 10). The self checking detects only arithmetic errors. It would not detect wrong assumptions, or inappropriate use of engineering model. The overview checking involves the review of a project with respect to assumptions, engineering model, etc. Detail arithmetic checking is not usually performed at this level. The independent checking is also called peer review, or parallel project checking. Another team of engineers is charged to design the same structure. These two teams may use different approaches, different assumptions, or different models. The objective is that with two different perspectives of the same project, independent checking reduces the chance that a wrong assumption is being made or a potential failure mode is being ignored or overlooked as in the case of Hartford Coliseum roof design.

12.5 Quantitative Analysis of Human Error

The discussion presented so far focuses on what has happened, what lessons one should learn, and what measures one can take to minimize the repeat of the same error. A number of studies have been performed and models have been proposed in the quantitative treatments of human error with respect to the reliability of the structures.

In 1986, a workshop on the modeling of human error in structural design and construction was held (ASCE, 1986). The workshop was sponsored by the National Science Foundation. Twenty seven researchers (22 were academicians) participated in the workshop with 17 papers presented. The topics presented include mathematical modeling and treatment of errors, error types observed, processes suggested minimizing errors. The workshop also included group discussions on error occurrences, structural reliability models, and error control strategies. A complete list of papers published and the reports of the group discussions is given in the Appendix.

The most comprehensive discussion on modeling of human error can be found in Melchers (1995). In that chapter, Melchers summarized some of the mathematical models developed on checking processes. Reliability analyses of human error can also be found in Melchers (1995) and Ellingwood (1987) as well as the publications resulted from the National Science Foundation sponsored Human Error workshop (ASCE, 1986). For readers interested in more in-depth discussions of research and studies done, they can refer to the suggested references in the Appendix. Both the Ellingwood (1987) and Melchers (1995) papers have many references cited on the topic of human error.

12.6 Conclusion

Four well known failure cases in engineering occurred during the years 1976 to 1986 are used to illustrate the consequence of human error. These cases were selected because the reports of the investigations are widely available. Some investigations are very extensive and performed independently (the investigators usually were appointed by a public agency instead of a party of the design or construction team). The four cases presented to, the Hyatt Regency Hotel walkway collapse in 1981; the Hartford Coliseum roof collapse in 1978; the Teton Dam failure in 1976; and the space shuttle *Challenger* accident in 1986, represent the types of human error that typically lead to unfavorable consequences.

Although the Hartford Coliseum roof collapse did not result in the loss of human lives, it was still a devastating failure. The casualties would have been very different if the collapse occurred a few hours earlier. The significance of these projects is that they all involved elements of unprecedented engineering, construction process, or condition where prior experience was not readily available. Similar situations can easily arise in tall building design and construction where engineers frequently face new understanding of the behavior of the building under various forms of loading or environment. The engineers also have to consider new design concepts as well as construction methods in order to accomplish the task economically and meet the clients' need without jeopardizing the integrity of the structure. The lessons learned from these four projects should help engineers be more conscientious of communications, responsibilities, and design in their next unique project as well as ordinary projects. The significance of communications, responsibilities, and design has been iterated in the wake of these failure cases. These factors are also discussed in Booth's chapter on "Quality Management of Structural Design" (Chapter 10).

The modeling of components of human error and its reliability formulation are mentioned but not fully discussed here. The author feels that it is more beneficial for the readers who wish to consider the risk of human error in their project more rigorously to consult the original publications for detailed discussion of the mathematical modeling and limitations.

Acknowledgements

The author would like to thank Mr. Glenn Cox, a former graduate assistant at the University of Tennessee, Knoxville for researching many of the articles used in this chapter.

References

ASCE (1986) *Modeling Human Error in Structural Design and Construction*, edited by A.S. Nowak, American Society of Civil Engineers, Reston, VA.

ASCE (1988) *Quality for Constructed Projects: a Guideline for owners, designers, and contractors*, American Society of Civil Engineers, Reston, VA.

ASCE (2000) *Quality in the Constructed Project*, 2nd edition, American Society of Civil Engineers, Reston, VA.

Blockley, D. (1977) "Analysis of Structural Failures", *Proceedings*, Institute of Civil Engineers, Part I, Vol. 62, pp.51-74.

BuRec (2002) U.S. Bureau of Reclamation Pacific Northwest Region website, http://www.pn.usbr.gov/dams/Teton.shtml, May.

Ellingwood, B.R. (1987) "Design and Construction Error Effects on Structural Reliability", *Journal of Structural Engineering*, ASCE, 113(2), May, pp.409-422.

Feld, J. and Carper, K.L. (1997) *Construction Failure*, 2nd edition, John Wiley & Sons, Inc.

Gillum, J.D. (2000) "The Engineer of Record and Design Responsibility", *Journal of Performance of Constructed Facilities*, ASCE, May, pp.67-70.

Kaminetzky, D. (1991) *Design and Construction Failures: Lessons From Forensic Investigations*, McGraw-Hill.

Luth, G.P. (2000) "Chronology and Context of the Hyatt Regency Collapse", *Journal of Performance of Constructed Facilities*, ASCE, May, pp.51-61.

Matousek, M. (1981) "A System for a Detailed Analysis of Structural Failures", *Structural Safety and Reliability*, Elsevier, Amsterdam, pp.535-544.

Melchers, R.E. (1995) "Human Errors and Structural Reliability", *Probabilistic Structural Mechanics Handbook Theory and Industrial Applications*, edited by C. Sundararajan, Chapman & Hall, pp. 211-237.

Moncarz, P.D. and Taylor, R.K. (2000) "Engineering Process Failure – Hyatt Walkway Collapse", *Journal of Performance of Constructed Facilities*, ASCE, May, pp.46-50.

Nowak A. and Carr R. (1985) "Sensitivity Analysis of Structural Errors", *Journal of Structural Engineering*, ASCE, 111(8), August, pp. 1734-1746.
Nowak, A. (1992) "Human Errors in Structures", *Safety and Reliability*, Vol. II, ASME, pp.335-341.

Pfrang, E.O. and Marshall, R. (1982) "Collapse of the Kansas City Hyatt Regency Walkways", *Civil Engineering*, ASCE, July, pp.65-68.

Rogers Commission (1986) *Report of the Presidential Commission on the Space Shuttle Challenger Accident*, June 6. U.S. Government Printing Office. Partial report can be obtained online through NASA Kennedy Space Center website, http://science.ksc.nasa.gov/shuttle/missions/51-l/docs/rogers-commission.

Ross, S.S. (1984) *Construction Disasters – Design Failures, Causes, and Prevention*, McGraw-Hill, New York.

Rubin, R.A. and Banick, A.M. (1987) "The Hyatt Regency Decision: One View", *Journal of Performance of Constructed Facilities*, ASCE, (1)3, August, pp.161-167.

Smith, E.A. and Epstein, H.I. (1980) "Hartford Coliseum Roof Collapse: Structural Collapse Sequence and Lessons Learned", *Civil Engineering*, ASCE, April, pp.59-62.

U.S. Dept. of the Interior (1977) *Failure of Teton Dam: A Report of Findings*, Teton Dam Failure Review Group Washington, D.C., U.S. Government Printing Office, 774 pp.

APPENDIX

This section lists publications where readers can learn more about other structure failures (hence the lessons learned) and the research done.

Construction Failures

Chapmann, J.C. (1998) "Collapse of Ramsgate Walkway", *Structural Engineer*, Institute of Structural Engineers, 76(1), January, pp.1-10.

Feld, J. and Carper, K.L. (1997) *Construction Failure*, 2nd edition, John Wiley & Sons, Inc., 512pp.

Hammond, R. (1956) *Engineering Structural Failures*, Philosophical Library, New York.

Kaminetzky, D. (1991) *Design and Construction Failures: Lessons From Forensic Investigations*, McGraw-Hill, 600pp.

Kletz, T. (1985) *What Went Wrong?* Gulf Publishing Company, Houston.

Kletz, T. (1993) *Lessons from Diaster,* Gulf Publishing Company, Houston.

Levy, M. and Salvadori, M. (1992) *Why Buildings Fall Down*, W.W. Norton, New York, 334pp.

Lewis, R.S. (1988) *Challenger: The Final Voyage*, Columbia University Press, New York.

McKaig, T.H. (1962) *Building Failures*, McGraw-Hill, Inc., New York.

Pugsley, A.G. (1966) *The Safety of Structures*, Edward Arnold Publishers Ltd., London.

Report of the Engineering Investigation Concerning the Cause of the Collapse of the Hartford Coliseum Space Truss Roof on January 18, 1978, (1978) Lev Zetlin Associates, Inc. (now the Thornton-Tomassetti Engineers), New York.

Report of the Royal Commission into the Failure of West Gate Bridge, Victoria (1971), Government Printer, Melborne, Australia.

Rogers Commission (1986) *Report of the Presidential Commission on the Space Shuttle Challenger Accident*, June 6. U.S. Government Printing Office. Partial report can be obtained online through NASA Kennedy Space Center website, http://science.ksc.nasa.gov/shuttle/missions/51-l/docs/rogers-commission.
NASA Actions to Implement Commission Recommendations is also available online

through
http://science.ksc.nasa.gov/shuttle/missions/51-l/docs/rogers-commission/NASA-actions.txt

Ross, S.S. (1984) *Construction Disasters – Design Failures, Causes, and Prevention*, McGraw-Hill, New York, 415pp.

Senders, J.W. and Moray, N.P. (1991) *Human Error: Cause, Prediction, and Reduction*, Lawrence Erlbaum Associates, Inc., Hillsdale.

Turner, B.A. (1978) *Man-Made Diasters*, Crane, Russak & Company, Inc., New York.

Vorley, G. and Lewis, J. (1998) "Ramsgate Walkway Disaster: a Wake Up Call for Certification Bodies", *Quality World*, July, pp.34-35.

NSF Sponsored Workshop on Modeling Human Error in Structural Design and Construction (1986) published by ASCE, edited by A.S. Nowak – papers presented in order of appearance:

Allen, D.E., "Human Error and Structural Practice".

Brown, C.B., "Incomplete Design Paradigms".

Ingles, O.G., "Where Should we look for Error Control".

Galambos, T.V., "H.E. in the Development of Structural Specifications"

Knoll F., "Checking Techniques".

Shibata, Heki, "The Role of Regulatory Documents in Controlling H.E."

Hadipriono, F.C. and Lin, C.C.S., "Errors in Construction".

FitzSimons, N., "On Modelling Errors in the Construction Industry".

Melgarejo, J.A. and Carr, R.I., "Decision Model for Human Error in Structures".

Stubbs, N., "Probability of Frame Failure Resulting from H.E.".

Blockley, D.I., "An A.I. Tool in the Control of Structural Safety".

Rackwitz, R., "Human Error in Design and Structural Failure".

Murzewski, J.W., "Formats and Responsibilities in Structural Design".

Ravindra, M.K., "Gross Errors, Seismic Margins and Seismic PRA's".

Lind, N.C., "Control of Human Error in Structures".

Turkstra, C.J., "The Case for Case Studies in Engineering Education".

Melchers, R.E. and Stewart, M.G., "Design Checking Models".

Frangopol, D.M., "Combining H.E. in Engineering Risk Assessment".

Grigoriu, M.D., "Effects of H.E. in Analysis".

Arafah, A.M. and Nowak, A.S., "Sensitivity Analysis for Structural Errors".

Hall, W.B., "Effects of Gross Error on Structural Reliability".

Pidgeon, N.F. and Turner, B.A., "H.E. and Socio-Technical System Failure"

Reports of "Working Group on Error Occurrences", "Working Group on Structural Reliability Models", and "Working Group on Error Control Strategies".

Mathematical Modeling and Analysis of Human Error

Balkey, J.P. and Phillips, J.H. (1993) "Using OSHA Process Safety Management Standard to Reduce Human Error", *Reliability and Risk in Pressure Vessels and Piping*, ASME, PVP-Vol. 251, pp.43-54

Ellingwood, B.R. (1987) "Design and Construction Error Effects on Structural Reliability", *Journal of Structural Engineering*, ASCE, 113(2), May, pp.409-422.

Melchers, R.E. (1995) "Human Errors and Structural Reliability", *Probabilistic Structural Mechanics Handbook Theory and Industrial Applications*, edited by C. Sundararajan, Chapman & Hall, pp. 211-237.

Nelson, W.R., Haney, L.N., and Ostrom, L.T. (1994) "Incorporating Human Error Analysis in System Design", *Safety Engineering and Risk Analysis*, ASME, SERA-Vol. 2, pp.155-159.

Stewart, M.G. (1993) "Structural Reliability and Error Control in Reinforced Concrete Design and Construction", *Structural Safety*, Vol. 12, pp. 277-292.

Chapter 13

Concluding Remarks

By:
Jun Kanda, University of Tokyo
Karen C. Chou, Minnesota State University, Mankato

This monograph has attempted to address issues associated with reliability-based design for tall buildings. It presents the fundamental concept of structural reliability analysis so that the users can perform the analysis for their designs, usually unique structural systems, or to expand further to meet their needs for the design analysis. The monograph also addresses issues associated with designing and constructing safe tall buildings that may not be described by mathematical models. These issues include durability in structural safety, quality assurance, quality management, quality controls in construction, and human factors.

Successful civil engineering projects cannot always be achieved through technical rigor alone. Tall building construction is no exception. Politics, economy, psychology, sociology, and other fields must collaborate to achieve and maintain safety. Even structural safety, as technical as it may appear, is influenced by the politics through the regulations, by economy through the market, by the psychology through the human error. Risk management may take place to enhance of the structural safety.

Design criteria to address events such as the World Trade Center collapse are still under heavy investigation and discussion. We may not know the recommendations and design impacts from these discussions for some time. As professionals in this field of structural design, especially for tall buildings, we have the duties to inform our clients of the possibility of reduced safety and its consequence even though the event may have a very small probability of occurring. Despite the event of September 11, 2001, tall building design will continue. Society still adores and admires the next tallest building coming up on the horizon. Thus it is our responsibility to continue our efforts for cultivating technology, knowledge, and imaginations for our future. We hope this monograph will offer you one important aspect of ensuring structural safety.

Index